I0803734

THE GREAT GOLD SWINDLE OF LUBEC, MAINE

CARRIE C. BANGS

EDITED BY
RONALD PESHA

THE History PRESS

Published by The History Press
Charleston, SC 29403
www.historypress.net

Copyright © 2013 by Lubec Historical Society
All rights reserved

First published 2013

Manufactured in the United States

ISBN 978.1.62619.008.5

Library of Congress CIP data applied for.

Notice: The information in this book is true and complete to the best of our knowledge. It is offered without guarantee on the part of the author or The History Press. The author and The History Press disclaim all liability in connection with the use of this book.

All rights reserved. No part of this book may be reproduced or transmitted in any form whatsoever without prior written permission from the publisher except in the case of brief quotations embodied in critical articles and reviews.

In Memoriam

In memory of Edith E. Comstock (1917–2012),
who lived all her life in North Lubec, collecting and preserving local history,
and was a contributor to this book.

In memory of Bernard M. Ross (1924–2013),
active Lubec Historical Society member for a third of a century and 1978
transcriber of the entire Carrie Bangs text from the original newspaper series.

Contents

Foreword

If you think about "gold from seawater," then we are living the modern-day version by residing in this small, seaside town of Lubec, Maine. Lubec is our treasure, perhaps even Maine's most curious little town you never heard of. We live in a place that draws its character from the water that surrounds us on three sides. We are lucky to have old salts living among us such as Ron Pesha. He spends his days tirelessly researching Lubec's past, including its true and very distinguished gold-from-seawater days of 1898. People believed in Lubec then, and even better, they believe in Lubec today. Our town brims once again with intrepid folks and entrepreneurial spirits. We hope you come see for yourselves because Lubec in 2013 is revived and ready for business. And unlike Jernegan and company, we know what an honest day's work is.

The Honorable Katherine Cassidy
Lubec, Maine Representative
126th Maine Legislature

Preface

The account of this remarkable 1898 hoax was chronicled weekly by the *Lubec Herald.* Half a century later, Lubec native Carrie Bangs penned a series of articles from 1949 to 1951 for that same newspaper. The opportunity to prepare her account, along with additions from numerous other sources (by no means comprehensive), fell to me.

My narrow and circumscribed research compels me to request submission of corrections and additional data. Considering the meager quantity of historical images herein, I urge anyone holding additional pertinent photographs (such as of the "iron bridge" at the Canal) to allow high-resolution scanning.

Acknowledgements

Contributors to this volume include Margaret "Peggy" and Harold Bailey, Pauline Bailey, Tom Dean, Patsy Kelley, Ginny McCaslin, Ginny Morano, Jennifer Multhopp, Davis Pike and Suzanne Plaut. Thank you to the Lubec Historical Society for the use of its documents and pictures; Edith Comstock for pictures; and Bernard Ross for transcribing the articles from Carrie Bangs. In addition, thank you to Gary Gree, the first president of the Lubec Historical Society, as well as all past and current members of the society for their diligence in collecting and preserving the artifacts, documents and articles on Lubec's history over the years. We are forever thankful.

CHAPTER 1
Initiating the Intrigue

Among the passengers debarking from the "Boston Boat" one evening in early 1898 were two distinguished-looking gentlemen who carried expensive luggage and exuded an air of influence nicely tempered by conservative good taste. One was obviously a man of the cloth, wearing the distinctive black suit and high-buttoned vest associated with the clergy of the period. Pushing through the crowd of passengers and onlookers, they beckoned to a carriage driver and directed him to transport them and their luggage to the Lubec Inn, the best hostelry in town.

Mr. Charles Fisher, formerly of Martha's Vineyard and until recently a floorwalker in a Brooklyn department store, and the Reverend Prescott Jernegan of Middletown, Massachusetts, had two things in common, despite a rather obvious disparity in their backgrounds. One of these was an aversion to gainful employment and the other a burning desire for a fast buck. Those who witnessed the pair's arrival could not know that they were seeing the opening act of a drama whose players were a cast of the most elite suckers ever assembled anywhere.[1]

Thus opens a brief account of extracting gold from seawater. This narrative expands greatly on the 1976 story, reprinting with additions and annotations a comprehensive newspaper series from the mid-twentieth century. We will learn background, the hatching of the scheme and its development and demise from Lubec native Carrie Comstock Bangs. Ms. Bangs (December 3, 1898–October 17, 1975) set out to write her full account of the swindle for the weekly *Lubec Herald* newspaper on her fiftieth birthday in 1948. Half a century had passed since the rise and fall of the scam. Bangs took advantage of 1898 issues of the *Herald* at hand for reference. The weekly periodical began in 1886, but the only known complete run from volume 1, number 1, through 1909 has since been lost. Thus, the sixty-three weeks of her articles preserve a primary source of history not otherwise available.

Bangs usually cites dates for her many quotes from the *Lubec Herald* and other newspapers such as the *Boston Globe*. Frustratingly, she also offers quotations lacking citations of sources or dates. Her columns are reproduced as written, each chapter a week's column. I offer occasional clarification and additional data within brackets.

THE GOLD FROM SEAWATER STORY OF 1898

Articles 1 & 2, December 16 and 23, 1948

The year 1948 marks the golden anniversary of Lubec's own "Klondike" of 1898. So much of this remarkable story has become so dimmed in the intervening years as to be almost forgotten. But with the help of those who remember, and by a study of the newspapers of the times, we may hope to recapture the essence of this fantastic tale.

Many absorbing things were going on that year: the Spanish-American War, Bismarck's death, the Dreyfus Affair and the real Klondike, or Alaska Gold Rush.

The Gold-from-Seawater story covers a number of states from Maine to Florida, and there were many people who were connected with or affected by this venture in some way.

Contradictions exist in many of the sources of information, due in part to reporters writing about an unfamiliar place from a distance. It will be our aim to give all possible angles that the reader may choose for himself the one that suits him best.

Before recounting the story of this magnificent fiasco, we might list the dramatis personae of the Electrolytic Marine Salts Company:

Reverend Prescott Ford Jernegan,
Vice-President and General Manager
Charles E. Fisher, Assistant Manager
Andrew N. Pierson, Superintendent
Arthur B. Ryan, President
William R. Usher, Treasurer
Albert P. Sawyer, Director
William Phelan, New York Detective
Marcus W. Jernegan, brother of Prescott F. Jernegan
Mr. Farmer, Bookkeeper
Big Jack Ives, Watchman
R.D. Shanahan, Contractor
William F. Arrington, Chemist
Workmen, Stockholders, and interested onlookers a-plenty.

The Klondike

Article 3, December 30, 1948

After a period of fifty years, only traces of piling remain to show that there ever was such a thing as "the project to extract from the ocean water, its gold."

In October 1897, two newcomers to Lubec leased from Hiram the old tidal water gristmill for purposes then unknown to the general public. This mill was a heterogeneous group of buildings that bridged the gap between Millard's Point and the land opposite, at the entrance of the Mill Pond at North Lubec.

About this time, Mr. Jernegan was heard to remark that millions of dollars were going through Lubec Narrows every day. That traces of gold do exist in ocean water was a generally recognized fact, and it was not unreasonable to believe that some method could be evolved for getting it out.

As far as Lubec is concerned, there were three leading characters in the organization that became

Klondike Plant No. 1 looking east. Originally this was the gristmill of Hiram Comstock, who later bought it back for his "Starvation" sardine cannery. The millpond is in the foreground. The structures, erected on pilings, spanned Mill Creek near its exit into Johnson Bay. Millard's Point is the left bank of Mill Creek. The house had been purchased by Reverend Jernegan and later burned, possibly in the August 6, 1921 fire that started on the west side of the Neck and swept across the main road. The *Lubec Herald* of August 11 headlined this event as the "Worst Forest Fire in History of Town." This picture was taken about 1910. *Courtesy of Edith Comstock.*

known as the Electrolytic Marine Salts Company. Reverend Prescott Ford Jernegan, a native of Martha's Vineyard and a minister of the First Baptist Church of Middletown, Connecticut, was the originator of the plan to procure the gold from the water. Mr. Jernegan lived in the house now occupied by Mrs. Dennis Tyler. At that time, thick dark spruce trees were plentiful near the house and the eastern end of the Mill Pond. This house was known as the Jernegan House for many years thereafter.

Mr. Charles E. Fisher, also of Martha's Vineyard, who had, among other things, been a floorwalker in a Brooklyn, New York department store, was locally regarded as the brains of the company. Mr. Fisher lived in lavish style in the

The Bangs House. Built on North Lubec Road after the marriage of Albert L. Bangs and Mary E. Comstock Bangs, the couple sold it to Andrew N. Pierson of Cromwell, Connecticut, a superintendent of the Klondike plant. Albert bought it back later in 1898, according to his son Olaf, married to Helen. The property was acquired by Captain James Bangs, Albert's father, in 1855 from Major Lemuel Trescott, who is buried in a small adjacent fenced military cemetery. It later burned. *Courtesy of the Lubec Memorial Library.*

house later owned by I.M. Bangs and more recently by C.H. Nugent.

Mr. Andrew N. Pierson, a florist of Cromwell, Connecticut, bought the house owned by A.L. Bangs [Albert Leon Bangs, 1865-1946], who later bought it back from Mr. Pierson.

It might be wondered why such a remote spot as North Lubec was selected for the work of this company until one recalls the unusual range of tides to which the Passamaquoddy region is subject. [This area lies within the lower Bay of Fundy, which experiences the widest tidal range in the world. Low to high tide at Lubec can exceed twenty-two feet within about six and a quarter hours.] Then, too, there was good boat service between Eastport and "the westward."

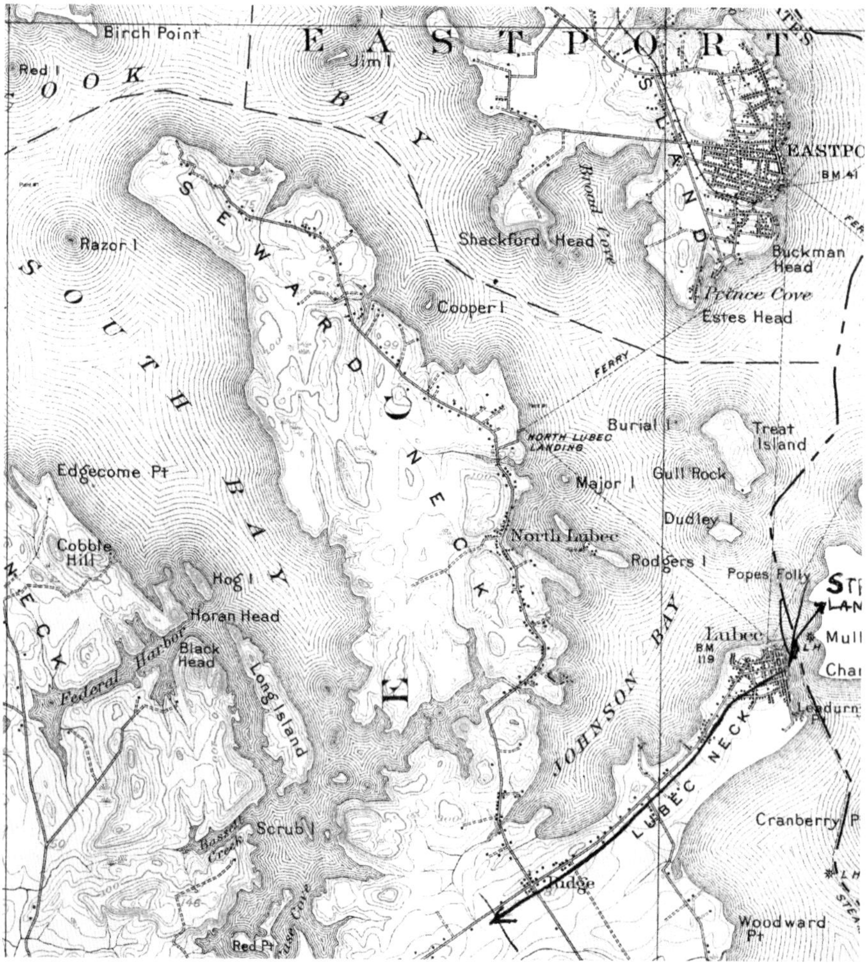

This 1914 U.S. Geological Survey map (Eastport Quadrangle) shows all of Seward Neck (North Lubec), as well as Lubec village at the northeast end of Lubec Neck. The site of North Lubec's Klondike Plant #1 is Mill Creek, just south of "North Lubec Landing," where the ferryboats arrived and departed. Farther south is a waterway crossing the Neck to South Bay on the west. Known as the Canal, the Electrolytic Marines Salts Co. began construction of Plant #2 by damming the entrance from Johnson's Bay on the east. *Geologic Atlas of the United States. Eastport Folio Maine.* Contour interval 20 feet.

The old "Starvation" or gristmill at Mill Creek, which had also seen service as a sardine factory, became known as "Plant No. 1." Gold began to be produced in paying quantities, and a few months later surveying and construction began on "Plant No. 2" at the "Canal."

People in general were skeptical of the idea of obtaining the gold at first, and men accustomed to large dealings in the stock market were doubtful of a concern whose promoters included a preacher, a floorwalker and a florist engaged in a work so foreign to their vocations. Lubec folks, too, were reluctant to place faith in the program, and the *Lubec Herald* of March 1, 1898, felt constrained to chide them for their slowness to accept the new order of things. The article said, in part, "Good fortune has befallen the Town…A company of able and wealthy men from Mass. and Conn. have located in our midst and are now operating the most wonderful plant in the known world. In spite of all ridicule and false reports the fact stands indisputable that at North Lubec gold and silver are being taken from the sea water in quantities that make the enterprise a paying investment…This is no visionary scheme. The property is bought and paid for, the survey is completed and within a few weeks the plans will all have been drawn and work be actually begun. The time for doubting and surmising has gone by. We must now awake to the realization of facts and act accordingly."

FIRST DREAM OF SEA'S WEALTH

Article 24, September 29, 1949

Prescott Ford Jernegan was born on December 17, 1866. That would make him about eighty three years

Prescott F. Jernegan. *Courtesy of Peggy Bailey.*

of age at present, if living, and there is reason to believe that he may be! [According to the website "Girl on a Whaleship," based on Jernegan's older sister Laura's diaries, Prescott Jernegan died at age seventy-four on February 23, 1942, while visiting Texas.]

So much is written in the newspapers of 1898 in regard to this story of the Electrolytic Marine Salts Company that should be made available to the people of Lubec, who, assuredly, have a right to all possible particulars. From the *Boston Globe* of August 7, 1898, appears an article that deals with the period when Mr. Jernegan was a student, and regarded as a "grind," at Boston [*sic*; should be Brown] University:

"It was undoubtedly during his college life that Jernegan's attention at first turned to his dreams of mineral wealth in and under the sea that afterward matured in the colossal swindle of The Electrolytic Marine Salts Company. In his graduating year about the time of the University Commencement, he made a visit to the works of The Sea Water Steel Company on the shores of Block Island. About that time, some summer visitors, reputed to be wealthy, had conceived the idea of gathering through some process the small particles of rich and valuable steel which washed up from the sea bottom along Crescent Beach [near Old Harbor, Block Island, Rhode Island].

"Several Brown University students were serving as waiters in the hotels of Block Island at that time, and they made some investigations on their own account. With magnets, they gathered a quantity of this steel and when they returned to college it became a matter of deep consideration among some of the old professors and students as well. Persons from the West, interested in the iron industry, established a plant on the sandy beach, just above the mark of the winter and storm tides.

"The place for half a mile was literally black with fine steel sand. The plant cost about $10,000 and consisted of an engine, a large tank or two and some machinery about which the projectors observed great secrecy. A considerable quantity of this valuable steel sand was obtained, through a system of magnets, at first, and later by a more complicated system which separated the white beach sand from the steel. The steel shipped away was found to be of a particularly valuable character, but there was not sufficient quantity washed up along the shore to make it a paying enterprise, and last year not a vestige of the plant remained on the beach. It had completely rotted away.

"It was stated at this time that during the investigations of the sea water steel projectors, some particles of gold had been obtained but not enough to make the effort to secure it remunerative. There are graduates of Brown University in Providence today who recall these incidents, and the interest that Jernegan took in the matter at that time. They believed that he at first considered the extraction of gold from sea water practicable, and that when he decided to stock his company he had an idea that with sufficient capital the thing could be made successful."

Now, of course, we know that steel is an alloy, or mixture of iron with some other metal or metals. Undoubtedly, that steel sand was magnetite, an angular and black iron compound common to the shore line of Long Island Sound. The chronicler has personally obtained some of this type of sand, near New Haven, with a magnet. But no gold, you may be certain.

Gold entered Jernegan's life some years later. Graduating from Brown University with ordination into the ministry, he taught for a year at Phillips Andover Academy. According to a published account in Edgartown, he read *Looking Back* by Edward Bellamy, a highly popular utopian novel of 1887 that initiated a Nationalist political movement. "This book changed the whole later course of my life," he said.[2]

Jernegan entered Newton Theological Seminary and then, in early 1892, accepted a pastorage at the Middletown Connecticut Baptist Church. His congregation, finding him a bit too liberal for its taste, tolerated him until July 1895. Finding himself in need of employment, his wife and small son settled in Boston when the president of Newton Seminary recommended him to DeLand Baptist Church in Florida in the Daytona Beach metropolitan area.

DeLand is home to the main campus of Stetson University, founded by John B. Stetson (1830–1906), who "invented" the traditional cowboy hat. Stetson was a major funder of Jernegan's new establishment, paying a third of his salary.[3] Stetson's religious ethics led him to offer good working conditions and benefits to his employees, a practice undoubtedly appealing to Jernegan's evolving social concerns.

Florida was extending the Atlantic Intracoastal Waterway, comprehensively chronicled in *Florida's Big Dig* by William G. Crawford Jr. A segment of canal development under the Walker Land Trust drew New England investors, including Newburyport, Massachusetts banker Albert P. Sawyer (1842–1903), who just happened to be an ardent Baptist. It is unclear if he met Reverend Jernegan at the Deland Baptist Church, but he somehow made propitious contact at a time when the land trust experienced cash flow problems. Seeking an influx of golden funds, Sawyer eventually became a director of Jernegan's company.

In DeLand, Jernegan also encountered his boyhood friend from Edgartown, Charles Fisher. Details of circumstances leading to this coincidence are not at hand. He also became acquainted with Albert P. Sawyer, a Newburyport, Massachusetts banker and Florida land developer who would become a director of Jernegan's gold plant in Lubec. "Sawyer and other promoters reportedly raised more than $5 million from Newburyport residents alone."[4]

No more endorsed by the DeLand congregation than at Middletown Baptist, Jernegan and Fisher left the city in 1896 and returned to Edgartown. Highly conflicting accounts attempt to explain the genesis of the gold scam, most revolving around Jernegan's visions and delirium:

> *By chance one day he read an article in the encyclopedia which told of there being gold in the sea. That was his downfall. Not being a scientific man the article appealed more to his imagination than to his sense and he thereupon began a career of fraud which has become classic not only in Connecticut but throughout the country.*

Another version of the story has it that Mr. Jernegan after attending a lecture that dealt with the enormous amount of gold there was suspended in the sea, had a most astounding spiritual experience:

> *On a train coming from Washington the worthy minister had reposed himself in his berth when in a burst of light the Lord appeared to him and gave into his keeping the secret of how gold could be taken from the sea. Mr. Jernegan having the mystery direct from heaven was not one to flaunt it in*

> *the faces of the uninspired scientists, but kept it locked in his own heart as all such revelations should be.*[5]

The secret of getting gold from seawater came to Jernegan, who had recently served several years as pastor of the Baptist church in Middletown, "in a heavenly vision while he was on his way to Florida, in a parlor car, in 1896," reported the *Middletown Penny Press* newspaper on November 15, 1897.[6]

A more detailed account appears in a publication of the then Dukes County Historical Society in Edgartown. During six weeks of typhoid, "his nurse read to him a newspaper story of the discovery of small amounts of gold in seawater by a noted English chemist," Edward Sonstadt, in 1872.[7] Meanwhile, his wife returned from a visit to Philadelphia and told Prescott she wanted a divorce:

> *It was a troubling year for Jernegan. A minister, he lost jobs with two churches, and his wife considered leaving him for another man. In his feverish delirium, he recalled an article about a recent discovery of trace amounts of gold in seawater. He wrote, "It instantly occurred to me that since gold was already contained in sea water it would be an easy matter to conceal more gold there, then to sell the 'secret' for a large sum of money and disappear into parts unknown." He saw the idea as a guaranteed way to "become something out of nothing."*
>
> *Jernegan contacted his childhood friend Charles Fisher, a charismatic and adventurous young man, to propose his money-making scheme. Fisher, also in financial trouble, enthusiastically agreed to join his friend. It would be easy, Jernegan figured, to exploit a rich, greedy capitalist who would not resist the riches of gold: "It seemed to me that I might justly plunder one of these men and, like the Robin Hood of old, use the money for good and generous purposes."*[8]

The extended account by Carrie Bangs states that "from the time that P.F. Jernegan had the vision about getting gold from sea water, while in the parlor car en route to Florida in 1896, a great many careful plans were laid before the Electrolytic Marine Salts Company came into being…In Florida it was that Mr. Jernegan and Charles Fisher 'went into cahoots' about the matter of 'extracting' the gold."

CHAPTER 2

Scamming the Suckers

For four months, Jernegan and Fisher tried to develop a method that would make their dreams of creating an oceanic gold mine a reality. Jernegan met with an acquaintance from his preaching days in Middletown, a wealthy jeweler named Arthur Ryan. He showed Ryan a small box containing a copper plate and a thin layer of mercury. This accumulator, he said, would collect gold as water ran through with the changing tide. An interested Ryan consulted with his friend, a florist named Andrew Pierson. The two men decided that they would invest if Jernegan and Fisher could prove that their device actually worked.

Unsure of how to proceed, Jernegan and Fisher went to Providence, Rhode Island, to perfect their demonstration. After several failed trials, they came across a method that involved adding an electric current to their mercury sample. Although this process did not actually collect gold, they believed the demonstration would be elaborate enough to fool their audience.

Jernegan met with the potential investors in Providence to prove once and for all that he could extract gold from the sea. To assure them that they were in complete control of the demonstration, Jernegan let Ryan and Pierson buy their own materials and conduct the experiment. The men gathered in a small hut on the pier and lowered a wired box containing a copper plate and an electronically charged layer of mercury a few feet into the ocean. The next morning, Ryan and Pierson took the contents of the box to a chemist of their choosing for analysis. They returned with astonishing results: five grains of gold and ten grains of silver were present.

Delighted, the two investors each loaned Jernegan over $5,000 to extract gold in mass quantities.

How did Jernegan and Fisher dupe their audience? The popular story told in newspapers was that Fisher, who did not attend the demonstration, wore a diving suit and inserted gold into the contraption while it was still under water. In his memoir, Jernegan insists that Fisher never did this. The partners entertained the idea but abandoned it after Fisher nearly drowned during a trial run. Instead, Fisher cleverly inserted gold into the mercury samples before they were assayed.[9]

Those Early Experiments

Article 34, August 4, 1949

From the time that P.F. Jernegan had the vision about getting gold from seawater while in the parlor car en route to Florida in 1896, a great many careful plans were laid before the Electrolytic Marine Salts Company came into being.

In Florida, it was that Mr. Jernegan and Charles Fisher "went into cahoots" about the matter of "extracting" the gold. New York seemed the logical city to get proper "contacts," and in September 1896, one William Phelan, a detective by profession, received a letter from Mr. Fisher, from DeLand, which referred to Mr. Jernegan and hinted at great things to be gained from a secret process of his in connection with the aid of electricity. No facts were divulged to Mr. Phelan at this time, but enormous future wealth was intimated.

Accordingly, Mr. Jernegan and Mr. Fisher went to New York, rented Mr. Phelan's front parlor and stayed a month. Many unique and peculiar machinations must have taken place, and next we find the two "partners" at Niantic, Connecticut, on the sound. Phelan also went to Niantic somewhat later, where Jernegan and Fisher were busy experimenting with chemical and electrical machinery. It appears that they were at the cottage of A.B. Ryan of Middletown, whom Mr. Jernegan had known while serving as pastor there. However, the residents of Niantic regarded the experiments and, perhaps, the experimenters with suspicion, and they were invited to leave.

Next, we find them in Providence where they were supposed to be making a geographical survey of the bottom of the bay. To get the configuration accurate, Fisher needed, he said, a trusted helper. So Mr. Phelan, who had gone to New York from Niantic, was asked to come and hold the life line for the important work to be done. Phelan, so we are told, took this in good faith and went to Providence with the promise of being paid well for his services.

On some diving expeditions, Mr. Fisher would be gone thirty five or forty minutes, according to reports of the time. Mr. Jernegan at the same time was busy running about, evasive as to the value of all this undersea survey.

Experimentalizing

Article 35, August 11, 1949

In Narragansett Bay, a mile or more below Providence, a long, narrow and rickety old wharf was located opposite a little, rocky island. This was the spot selected by Jernegan and Fisher to try the experiment that would convince certain "capitalists" that the plan to extract gold from seawater was a feasible one.

On this wharf, a small, loosely built house was constructed for Jernegan's use in conducting his "survey." Soundings from the island to the wharf showed the water to be from seven to nine feet deep.

There are not many feet difference in the depth of tides at that part of the coast.

An insulated wire, or cable, was laid along the bottom connecting the island and the wharf. By this device, Fisher was able to follow along the bottom in his diver's suit without the aid of Phelan to hold the life line. Instead, Fisher carried a tank of compressed air on his back and was able to make the trip from the island to the wharf and back in about fifteen minutes. While under the wharf, he could manipulate the machinery in the apparatus that was suspended beneath the house of boards.

Phelan remained on the island during this practice and is reported to have asked, "What are you up to anyway?" "We are going to make gold," was Fisher's reply at this juncture, although no details were given at this time. It is said that Phelan, true to his detective's nature, took notes during these experiments.

He was told that they were expecting two men from Connecticut to view the results of the experiments, with the idea in mind that if they were impressed, they would advance money for the enterprise. Fisher said that the capitalists would furnish their own chemicals and place them in the water themselves while Jernegan would supply the battery connecting with the submarine receptacle for getting the gold from the sea. Platinum wires were to be connected to the apparatus, which would make it easier to attract the gold that was in solution.

In due time, one cold day in February, Mr. A.B. Ryan and Mr. A.N. Pierson arrived in Providence, registered at the City Hotel and proceeded to the wharf to spend the night.

The Experiment on the Wharf

Article 36, August 18, 1949

There exists a sketch of that long, narrow and rickety wharf, which was located below Providence, where the demonstration was made to prove that gold could, with proper equipment, be taken from salt water. This drawing appeared in the *Boston Sunday Herald* of July 31, 1898, and was entitled, "How the Scheme Worked."

A blunt headland shows as a background, and the surface of the Bay appears pacific–let us say, to be funny. This shows that the artist was operating under the influence of the torrid weather common to July and had suffered a lapse of memory that the experiment had been performed in February "with the ice cakes bobbing around." Nonetheless, the picture gives the idea.

An ill-constructed and too well-ventilated "house of boards" is shown on the end of the wharf wherein five men gathered around a winch. Two of the gentlemen are seated; one, presumably representing Mr. Jernegan, is explaining while the other four men are looking on interestedly. According to accounts of that story, only Jernegan, [Andrew N.] Pierson and [Arthur B.] Ryan were in that demonstration. Maybe this represents a preliminary showing. One man is attired in a dress suit and tall silk hat. The others are sporting derbies.

Beneath the shack, pussyfooting along to a large square box suspended from the wharf, goes Fisher. Clad in his diving suit with a long cylindrical can of compressed air on his back, he has followed a wire from a small rocky island to the submerged box. The water there is several feet deep, and he is well covered. With the brisk winds and darksome waves of a cold winter's night, there was small chance, indeed, that he would be seen or heard. Four ropes,

Drawing from a mid-1898 edition of the *Boston Sunday Herald.*

or chains, one from each corner of the box meet in the center where they are fastened to the main rope controlled by the hoisting device. We shall next try to find out what was said and done that freezing February night of 1897.

THE NIGHT ON THE WHARF

Article 37, August 25, 1949

There exist speculations and various accounts of what took place in that shack on the end of the long, narrow wharf below Providence Harbor on the cold February night in 1897. Practice had been successful where Fisher was able to travel under water by following a wire cable that led to the gold-producing box beneath the wharf.

All was ready, and Mr. Jernegan was supposed to have been the one to meet the capitalists from Connecticut. These men, Arthur B. Ryan of Middletown and Andrew N. Pierson of Cromwell, alighted from the train at Providence and registered at the City Hotel. From there, they proceeded to the wharf where the nature of the device was explained to them. The gentlemen had brought their own quicksilver with them, and one account states that Mr. Ryan had an appliance made according to Jernegan's directions, namely "a gold accumulator," a flat box no larger than a plate. It contained a battery, and into this box the mercury and other chemicals could be poured. Arranged so that the salt water could flow over these materials, when the switch was turned on, the extracting process was on the way.

How bitterly cold that night was; how drafty the shack. What warmth there was, was furnished by an oil stove. But possibly there was added warmth in the thought that the success of the experiment would yield untold wealth for them all in the future.

Using a crude winch, a log and crank, they were able to lower and raise the larger box that held the mechanism. Boards that were bolted to the floor were opened, and the men put in the chemicals to which Jernegan added something from a vial and lowered away. The switch was thrown, and the platinum wire that connected the wharf and the box was supposed to be the best type of wire to make the experiment work effectively.

Some say that Jernegan, at this point, betook himself to his hotel to spend the night in comfort, leaving Ryan and Pierson alone in the "house of boards." Other accounts say that the three men stayed and probably shivered through the night.

The Morning Following

Article 38, September 1, 1949

During the course of the night, while Mr. Ryan and Mr. Pierson, city-dressed men, shivered in the eight- by ten-foot "house of boards" at the end of the wharf, Fisher, unobserved, began to play his part. Donning his diving suit and the tank of compressed air, he made his way, unaided, out along the piling of the wharf to the "accumulator."

"We use no power, you know," Jernegan had said, "and we have to operate the works on the tide water system. You cannot operate a plant of this kind at will." Thus, some explanation had to be offered for the unusual time of the experiment. Where the gentlemen from Connecticut had arrived in the afternoon, they had presumably missed the day tide.

The darkness of the night, the noise of ice cakes and the wash of the waves aided Fisher while he dumped out the existing and "honest" chemicals and replaced them with his own solution of mercury with gold.

On his arrival the next morning, Jernegan pulled up the box with the idea that they were to get the gold "before their very eyes." The thought that "one picture is worth ten thousand words" must have been part of Jernegan's philosophy.

It seems that Mr. Jernegan's pulse skipped a few beats when he saw that a portion of the mercury had leaked out, characteristic of its nature, through the lead lining. That would leave about an estimated $2.00 worth of gold content to be assayed. It appears, at this point, that he made some joke of the matter and slipped into the solution a small California nugget, which he wore as a stick pin ornament. This added to the yield, to be sure, and when an assayer had tested the gold-bearing quicksilver, it was found to be valued at $4.50. The capitalists were impressed and amazed and expressed themselves as willing to put up money, on a large scale, for the buying of machinery. They departed from Providence with glowing enthusiasm.

The above description is adapted from the exposé written by Detective Phelan. [Bangs fails to cite the source of this document.]

Months later than that cold February night, in fact, on August 1, 1898, after the collapse of the E.M.S. Co., President Ryan gave out this statement: "I believe Phelan's story is correct. I was one of the men referred to who made that test on Providence Bay in February 1897, and I shall not soon forget the experience. My companion was a Connecticut man named Pierson. It was a bitter cold night, and we spent the whole of it in that little shanty at the end of the wharf. I remember that great cakes of ice were floating about, and we could hear them crunching against the piles.

"We had a couple of oil stoves burning. We stuffed the cracks of the wharf with paper and kept a-dancing, but we did not experience much comfort. I know nobody entered our hut during all the time we were there, and of course, it never occurred to either of us that somebody was creeping along under us, under those floating cakes of ice, and coolly dumping the gold into our box. As a matter of fact, Fisher was the last person that I should have expected, as it was represented to me at the time that he was in Florida, and I had not met with him again until some weeks later.

"When we found gold in the box, next morning, we were convinced that the scheme was all right. We furnished our own chemicals and no one had interfered."

Another account of the experiment appeared in the *Hartford Courant* on Sunday, January 17, 1926. The paper devotes the entire front page of its Fifth Section to the Jernegan scheme of three decades past, placing the site of the rickety wharf elsewhere. The newspaper text follows here:

Dredging Gold from Seawater

[Jernegan] *enlisted the services of Charles E. Fisher, a many sided genius who was not only adept at deep sea diving, but was an expert floor walker in the Brooklyn department store of A.I. Namm & Co., when Mr. Jernegan discovered him.* [Namm opened its store in 1891 at 452 Fulton. Becoming Namm-Loeser in 1952, the Fulton Street store closed in 1957, and a Bay Shore branch was sold to Gimbel's.]

Together they erected a shack in Niantic [East Lyme, Connecticut, less than twenty miles southeast of Middletown and near the U.S. Coast Guard Academy at New London] *and started experiments on their scheme. William Phelan, a detective from New York, was "let in" on the scheme by Fisher. It was through Phelan that the story finally leaked out. For some reason, Fisher and Mr. Jernegan removed to Rhode Island, and there in Narragansett Bay they hired a pier upon which they built a three-sided shack with a trapdoor that opened down into nine feet of water.*

Mr. Jernegan's reputation for honesty in Middletown led him to search for his first victims there. A.N. Pierson, the owner of the great Pierson Greenhouses, and A.B. Ryan, who later was made president, were talked into a semi interest in this gold project, and when all was in readiness, they were invited to Rhode Island to see Reverend Jernegen's "secret process" [to] *bring gold out of the ocean.* [An early Raphael Tuck & Sons postcard shows a vast area of ten or twelve greenhouses, some up to one hundred feet long, in a photograph taken from a nearby hill. The "undivided back" postcard indicates its date as 1901–7.]

The little party assembled on the pier on a bitterly cold winter's night. The financiers brought their own chemicals as Mr. Jernegan pleaded poverty. On one side of the shack was a most impressive-looking piece of apparatus, which later turned out to be only a pile of junk. A battery and a skillfully constructed box suspended on a windlass were the only other properties in the little drama.

The Daybreak Experiment

Fisher was not present. He and Phelan were on the mainland, where Fisher had donned a diver's suit "in order to make a scientific investigation." Phelan held the life line. Fisher salted the box as soon as it was let down;

> *into the mass of different chemicals he put three or four grades of gold and a little nugget having come out of Jernegan's scarf pin.*

Most accounts of the Klondike hoax refer to Charles Fisher as a skilled diver. This was not so, according to the article "Charlie Fisher, the Perfect Partner," published by the Dukes County Historical Society (since 2006 the Martha's Vineyard Museum, Inc., of Edgartown, Massachusetts) in the *Dukes County Intelligencer*. Use of diving apparatus to place flaked gold on the submerged electrodes had been planned, but during a test run, Fisher almost drowned. "Hauled ashore and resuscitated by Prescott, Charlie threw the diving suit into the ocean and never used it again." Other means, described later, were used to suggest that mercury in the "accumulators" amalgamated with the seawater gold.[10]

> *The box might have been pulled up five minutes after it had been let down. But that was not the way the reverend gentleman worked. He was a showman. He waited until dawn before pulling up the box. There was the gold. The Middletown men had been there and seen the whole thing. There could be no chance of their being hoodwinked. The chemicals were worth $4.50. What a profit!*
>
> *In ten days the first company was launched with a capitalization of $250,000. Plans were under way to make this a million. The town of Lynn supplied $200,000. New York and Boston gave their share, but it was his native* [sic] *Middletown that Jernegan found his happiest hunting ground. Every Swedish person in the city earned, or borrowed money to give Mr. Jernegan.*

Myron H. Avery (1899–1952), a native of North Lubec, wrote a related note. Avery, a navy veteran of both world wars, graduated from Bowdoin College and Harvard Law School and practiced admiralty law. He also became a major figure in development of the Appalachian Trail and the first to walk its entire two-thousand-mile length. His multiple interests extended to the history of North Lubec.

Avery's Lubec-related papers include an undated single-page document, apparently in his hand, which disputes the Narragansett Rhode Island site and moves the location thirty miles west, back along the coast to Connecticut, but not to Niantic:

> Discussed this article [in the *Hartford Courant*] *with Robert Booth Senior of Hartford...His father-in-law was familiar with these events. Never heard of William Phelan. Instead of Narragansett Bay, the experiment took place off of the government pier in New London* [Connecticut] *Harbor. A diver from Lott Wrecking Company was employed. Pierson's fortune is too large in this article. He probably contributed $5,000 and his employees $10,000. It is thought by Mr. Booth's father-in-law that $30,000 was invested in all by inhabitants of Cromwell* [between Middletown and Hartford, Connecticut]. *Many mortgaged their farms. Mr. Pierson had occasional letters from Jernegan after the scheme collapsed.*[11]

Regardless of the demonstration's venue, Jernegan and Fisher, or whoever the diver was, made a convincing pitch, leaving Jernegan poised to build his magnum opus in Lubec. Yet the diver legend persisted. The following article, written by Bill Aye, perhaps a pen name, appeared in the *Lubec Herald* on April 18, 1946:

> "KLONDIKE" FOLDED UP WHEN INVESTIGATORS FOUND DIVING SUIT
> *They had built a wharf* [at North Lubec], *with places at the bottom to hold the boxes that were to extract the gold. These boxes were very crude on the outside, being built of two-inch rough plank about three feet long by two feet wide with spaces of one-inch wide to let the sea water through. There was an inside container of wood, a little more finished, and inside of this the real gold extractor, which was a metal box with a compartment to hold the secret chemical that would separate the gold from the sea water that flowed through.*
>
> *Now, as to the test, as these boxes were set with the proper people there to see that there could be no chance of anything being dishonest, the chemicals were placed within the inner box and then the witnesses saw the tide cover them. Watchmen were posted so that no one could come near the wharf until the proper time would come to open them. Now comes the secret of the "gold from the sea."*
>
> *On a little island called Major Island, about one hundred and eight feet from the boxes that were being carefully guarded* [just outside the mouth to Mill Creek], *Mr. Fisher donned a diving suit and Mr. Jernegan manned the air pump. Mr. Fisher in his diving suit then walked over to the boxes and opened them and deposited therein a selected amount of gold nugget that had been previously prepared for the occasion, so nicely prepared*

as to look as though they had been chemically acted on, to not only astonish the select few that were allowed to be present, but to send wild stories to the press (and there was a press agent present) of the wonderful success of obtaining gold from sea water by the secret method of the Electrolytic Marine Salts Co.

CHAPTER 3

First Factory Fabrication

An article in the *Lubec Herald* of July 21, 1927, by the man who sold property to Jernegan opens this chapter, after which we return to the Carrie Bangs series:

> THE MILL TRANSACTION FOR PLANT NO. 1
> *The old Comstock tide mill built at North Lubec in 1820 by Robert Tidd* [sic] *is being torn down to salvage the sound timbers and the refuse is to be gathered up for firewood. For many years this mill did the grinding for all of Passamaquoddy, its output for a year sometimes reaching 30,000 bushels of corn. It also ground wheat to the extent that it was then raised hereabouts.*
>
> *With the coming of Jernegan and his Scandinavian dupes, the beginning of the end came for the old mill. Jernegan, who was an ordained minister, came to the writer one Sunday in October* [1897] *and said, "If it wasn't Sunday, I would give you so much for the mill property." Declining at first to sell I finally agreed to meet him at the Hotel in the village, where a trade for a modest amount was made. Then began the disemboweling of the old mill.*
>
> *A big crew started excavating and there were built long flumes of hard pine, with short spouts running from them into the accumulators, metal boxes which contained quicksilver, placed there to gather the gold from sea water. Great secrecy was maintained here. A high board fence was tipped out with several strands of barbed wire. An armed guard paced the wharf on the inside of the fence, day and night. I might say that the selection of*

This drawing's original caption reads, "Site of the New Plant Soon to be Constructed for the Electrolytic Marine Salts Co., at North Lubec," reproduced in the 1976 book *200 Years of Lubec History, 1776–1976. Courtesy of the Lubec Historical Society.*

these men was a good one, for had the fake continued until now, they would have still considered it genuine.

As the water passed through the flume under the wharf it was taken up by the boxes, and each weekend saw several hundred dollars worth of metal taken out. Was it any wonder that the public went wild? As fast as a new block of stock appeared, investors bobbed up from everywhere. Every western steamer brought more of the curious. I have seen the Mill bridge black with people hoping to gain some of the secret of the "goings-on" and I have been offered bribes by the more interested ones to divulge some of the secrets which, alas, I did not possess.

Extensive Plans of the E.M.S. Co.

Article 10, February 17, 1949

Previous to locating at North Lubec, other possible sites had been considered by those most concerned in the operation of the gold-extracting machinery. As we have seen, the gristmill at the Mill Pond was selected. The old grain-grinding machinery was removed from the mill and other machinery put in its place. By the spring of 1898, the company expected to have 220 machines in this inlet. It was asserted that each machine was capable of extracting one dollar's worth of gold and a quantity of silver in twenty four hours. Thus the 220 machines would capture $220 in gold and enough silver to pay all expenses. This item from the *Herald* gives us an idea of the wages paid during the very first months of the E.M.S. Co's sojourn. "During the building of the small plant at the inlet, the company paid out in wages each week, more than $400." And who shall say that four hundred dollars in the winter of 1898 was not welcome?

The company trod cautiously. Even in February the *Herald* stated, "As yet, the company has not taken a large amount of gold, but we have been assured on reliable authority that Mr. Jernegan has, in a recent experiment, secured $3.00 in gold from a single machine in one day." In such manner does gossip travel.

After a time it was announced that they were obtaining gold in paying quantities at the old mill. Of course, the stock went up, and many thousands of dollars' worth of shares were sold. The word got around, and from the March 22, 1898 edition of the *Lubec Herald*, an independent local newspaper of today published by "R.G. & F.L. Getchell Editors and Proprietors," came a preview of the visiting activities that were to follow.

"Our new neighbors at North Lubec Klondike are creating some excitement in business circles to the westward…Recently, the Electrolytic Marine Salts Company issued from their Boston Office three hundred and fifty thousand shares at one dollar each and in three days disposed of nearly the entire block. The moneyed men of Massachusetts are taking great interest in this new enterprise and are ready to take shares to almost any amount.

"Messrs. Louis Brockway and Jas. G. Foster, western capitalists, arrived by last Thursday's boat to investigate the new El Dorado situated at North Lubec. These two gentlemen were here three days and visited the Electrolytic Marine Salts Company works, where they were shown the gold and silver being refined in the laboratory. They returned to Boston Thursday fully convinced of the success of this wonderful discovery of Mr. P.F. Jernegan. These gentlemen will probably return later in the season to pass the summer at North Lubec. They express themselves as much pleased with the courtesies extended them by Mr. Fisher, assistant manager of the Comstock Plant."

After Plant No. 1 was well established, it was decided to enlarge the enterprise. Partly to find a suitable location, and partly as a blind, the E.M.S. Co. bought land including Elic's Creek clear up to South Bay. This, they said, would be in reserve for future developments and to keep out any possible rival companies.

As part of the project, Mr. George Mowry did surveying as far afield as Pembroke. There were rumors, well founded ones, of capitalists from Worcester and from Springfield, Massachusetts, asking to be partners in this enterprise, and deals were made. However, work did start in earnest at the Canal in the spring of 1898 [see Chapter 4].

Lubec, you see, had this bonanza thrust upon it, and willy-nilly, had to believe in it. But we rejoice to say, the municipal palms of Eastport also began to tingle according to a write-up for May 17, 1898.

"Indications point to the purchase of the property on the south side of the carrying-place–two miles out island–where a plant will be erected by the company for the purpose of extracting gold from the salt water in the Passamaquoddy Bay, as is now done at North Lubec. The Electrolytic Marine Salts Company who own and operate their "gold factory" across the bay from the above property, is now negotiating for the purchase of this very desirable site, where a quantity of sea water rushes in and out over the flats twice a day. A breakwater will be built from

Above: Klondike Plant No. 1, circa 1930, after its original owner, Hiram Comstock, bought his mill back and converted to canning sardines. *Courtesy of the Lubec Historical Society.*

Left: This drawing's original caption reads, "So Much Has Been Done on the Office for the New Plant," reproduced in the 1976 book *200 Years of Lubec History, 1776–1976. Courtesy of the Lubec Historical Society.*

This photo from December 4, 2012, shows the site of Hiram Comstock's gristmill spanning Mill Creek. It became Klondike Plant No. 1 in 1898. *Photograph by Ronald Pesha.*

the shore at the entrance of the basin and a large building erected if the arrangements can be made satisfactory with the owners of the property. With a gold factory in operation in the Island city, in connection with the many sardine factories, Eastport should make quite a rush for first place in Eastern Maine later in the year. With quantities of gold in the salt water, as claimed, there is little need of a trip to Alaska in the future."

It did, indeed, seem less arduous to get the gold from the water than from the Alaskan fields, and as someone later pointed out, it was even more easy to pick the gold from the pockets of the stockholders than from either of these places.

The Accumulators

Article 6, January 20, 1949

Beneath the "machine room" and the "laboratory" of the old gristmill at the Mill Pond at North Lubec were placed specially constructed boxes, called accumulators, for collecting the gold content of the salt water. These boxes were probably not so very sizeable, although they were, undoubtedly, accumulators of different specifications as experiments went on. These boxes were made in part of copper and containing a battery, mercury and unknown chemicals. It is recorded that one of the first accumulators that was used for demonstration

Simulated accumulator, assembled by Bernard Ross of the Lubec Historical Society, showing platinum rods in an iron pot, submerged under water at high tide and exposed during low tide. *Photograph by Ronald Pesha.*

purposes was lined with lead and was not much larger than a plate. The lead lining proved to be a bad idea, as the mercury could easily eat its way through this metal.

Just before February 15, 1898, there were one hundred of these "machines" under the wharf. Each was separate from the others. Officials then said that there was room for two hundred twenty machines in all and that they would soon be put in.

The accumulators were always under water, for the company had constructed at Plant No. 1 automatic tide gates, which would close at the beginning of the ebb tide and impound the water of the Mill Pond. There must have been gates at both the eastern and the western ends of the buildings. It may be that some of the salt water seeped out slowly during the considerable time that the tide was away from this "New Klondike." Was there ever a more appropriate name for a venture or a more obliging coincidence?

Between October 1897 and February 1898, some fifty or sixty men had been employed transforming the gristmill. The buildings were then protected from intrusion by a high board fence with three strands of barbed wire at the top and a gate bearing the inscription "No Admittance."

According to the June 28, 1898 edition of the *Lubec Herald*, "There are working at present 239 machines producing $239 a day net profit."

Charles E. Fisher is the one generally credited with "salting" the boxes or putting the gold inside. He was a diver and a good one and doubtless "worked" in the dead of night. Many think that he had at least one trusted accomplice to do some of this "depositing." Mr. Fisher, it was, who always designated which of the accumulators should be examined. He alone was responsible for the knowledge of the chemicals that went into them. Had this "secret" not been guarded well, any chemist would have known what to do to make the ocean yield its gold, it was averred. [Fisher of course did not place gold within the accumulators but exhibited gold ingots salvaged from other sources.]

There are many versions of how often the accumulators were examined. Some say daily, others say weekly and yet others say monthly. A plausible account describing the operation during July 1898 says that the inlet to the Mill Pond accommodated 240 accumulators, of which sixty were pulled up each week. Thus, each box was under water a month before its turn came to be examined. During that time, the water, chemicals and electricity had time to "work this marvel." It is known that Mr. Fisher was sometimes away for a few days at a time. His presence would not be necessary constantly if the boxes were emptied once a week.

Mr. James D. Huckins recalls one time that there was not so good a yield in the accumulators as expected. Moreover, some of the many interested investors from "away" were present and demanded an explanation. Mr. Fisher was hard-pressed for an answer when "Munt" Reynolds said, "Think of all this heavy rain we have had lately. All that fresh water running into the Mill Pond has diluted the salt water so that there couldn't be so much gold as usual." This saved the day for Mr. Fisher, and he echoed gratefully, "Of course that's it. This farmer knows what he is talking about." So Mr. Reynolds and the fresh water brook deferred the day of reckoning.

FURTHER ACTIVITIES OF THE ELECTROLYTIC MARINE SALTS CO.

Article 9, February 10, 1949

The president of the E.M.S. Co. of 1898 was Arthur B. Ryan of Middletown, Connecticut. Mr. Ryan was a jeweler by profession and one who, like Archimedes, by virtue of his knowledge of various tests for metals, certainly would know whether a substance was genuine. [The Middlesex County Historical Society reports that the Middletown City Directory for 1898 lists Arthur B. Ryan, 241 Court Street, of Ryan & Parker "watches, jewelry etc.," 226 Main Street.]

Being convinced that the metal produced by the Jernegan process truly was gold, and having confidence in Mr. Jernegan as his pastor, Mr. Ryan entered enthusiastically into the promotion of this plausible enterprise. Mr. Ryan, it was, who, in talking with the editor of the *Lubec Herald* in February 1898, stated, "You, Mr. Editor, are the second newspaper man who has asked for information. Yes, we have expended some $48,000, and may put out more, but are not fully decided. We have looked over half a dozen other places about Passamaquoddy Bay, and may decide to locate at one or more of these places, in preference to building at the Canal."

It appears that although, to quote the *Herald*, "much has been said and written about this project for extracting gold from sea water, many adverse criticisms have been made. Skeptics have sneered at the idea of gold being taken from salt water. The editors of various New England papers have written lengthy articles on the subject, letting their imaginations run wild. But only one newspaper man, besides the writer, has had a personal interview with the president and general manager of the company. We found both Mr. Jernegan and Mr. A.B. Ryan, president of the company, to be most genial and courteous gentlemen. These officers assured us that they were always glad to welcome any and all newspaper men who came to them for information, but that they did resent some of the articles in the daily press, which articles were without foundation, and for which the writers had never asked an interview." So there you have it.

North Lubec Gold Factory. Reproduced in the 1976 book *200 Years of Lubec History, 1776–1976. Courtesy of the Lubec Historical Society.*

At this time, there was much talk about having a branch railroad extended to Lubec. It actually looked as if the railroad might materialize. On commenting about things in general, the *Lubec Herald* of March 1, 1898, observed, "But a greater piece of good fortune has befallen the town than as though the 'Shore Line' were being constructed through here." This was in reference to the E.M.S. Co., of course, and there is a revealing item in the paper of Tuesday, June 14, 1898 (now someone check on June 14, 1898, to see if it was a Tuesday). [It was.] In passing, let us recall that the *Herald* has not always been published on Thursday. The item says: "Eight thousand dollars a month is being paid to the employees here by the E.M.S. Co. The railroad is not to be compared with our new industry."

Not everyone realizes that the first telephone line to North Lubec was due to the efforts of this remarkable company. The *Herald* of April 19, 1898, made this announcement. "The E.M.S. Co. wants a telephone connected with our village. It is needless to say that they will have it as what this company takes hold of goes. It will be a great convenience not only to North Lubec but to us in the village as well, and is another sign of progress. Electric lights and waterworks next." We believe that Lubec did get its "water works" in 1903. [The town began providing public water in 1901, derived from a copious spring of pure water in South Lubec. The same site continues as the municipal water source in 2013.]

By now, you have all had a good laugh, it is hoped, but to pursue the telephone project further, it seems that it was a worthy precursor to the 115 line of some years back that had many more subscribers than was good for it.

An item in the edition of May 3, 1898, gives the following opinions. "That there is to be a telephone line running between North Lubec and the village is an assured fact. There are twelve residences and places of business at the former place in which a telephone will be placed. There has been some delay in getting a representative of one of the New England Telephone companies down here. If he does appear soon Mr. Chas. Fisher proposes to put in a line, supplying the money himself. The fact that Mr. Fisher is interested in this assures the "success of the plan, for whatever the people of the E.M.S. Co. want is sure to materialize."

What must have been the thoughts of Messrs. Jernegan and Fisher as they allowed all these projects to go on. They evidently did not mind in the least what direction the ideas of Mr. Ryan and Mr. Pierson took. Perhaps they even had begun to believe in the scheme themselves. Mr. Ryan, Mr. Pierson and others were working assiduously making extensive plans for the future, commensurate with the vast amount of water to be used and the resultant gold to be reclaimed.

And they were working in good faith. With visible evidence of the good works already mentioned, what Lubecker could be blamed for being deeply interested, even believing? Small wonder, the *Herald* was impelled to point out. "The presence of these people is not only desirable for the amount of money they will bring into the town, but we should welcome them for their social qualities. The officers of the Company are earnest Christian gentlemen, and many of their employees are Christians. They have shown themselves broad-minded and liberal in their business relations. We wish them all success in their undertaking, and hope that they will take millions of dollars from old Passamaquoddy Bay…and we believe they will."

(Chronicler's note: At this point, the recorder of this preternatural tale confesses to becoming hysterical.)

NUGGETS FROM THE CANAL

Article 12, March 3, 1949

The column with this significant title began in the spring of 1898 and usually was written and punctuated as "Nuggets from the 'Canal.'"

It seems a pity to omit any of the items of that spring and summer of remarkable doings. Only seven of these weekly columns are available at present. Some of the items may have us guessing, as in the case of the following for May 31. "The bronoho [*sic*?] and Willis had the rheumatism last week and did not work Friday and Saturday. It is said that there is still one spoke left in the road cart and Willis will travel on this as long as it lasts." Still, someone may recall just what it is all about. Whatever it was, it doesn't sound very efficient.

The next item is in regard to that busy tug mentioned last week, and which evidently operated "for the duration." "The Tug *Sea King* brought down two large rafts of logs Friday. She started with four but the chains parted and she lost two of them."

At this same time, Mr. George Mowry was doing towing for the E.M.S. Co. with the Mizpah.

Out-of-state papers noted that board was very high, as well as somewhat hard to get, at this busy period. In addition to those who came for employment, and had to be housed, there were many visitors. The following will serve as a sample. "Capt. Leander Morton has a house full of boarders, all working for the E.M.S. Co. among them are Messrs. Merrill, Nason, and Bachelder, civil engineers, Mr. W.H. Stevens of Conn., Mr. Lynch of Florida, Mr. Aston and son of New York, Mr. H. Bidwell, timekeeper for the E.M.S. Co., Mr. Gannon, Mr. Alex McInnis, and F.W. Marshall."

In the fragments of newspapers at hand, no food prices were mentioned. Dry goods only were advertised. Probably the board was all of four dollars a week.

In the *Herald* of June 7, 1898, we read that "Mr. Shanahan returned from the West Thursday." R.D. Shanahan was the contractor for the work at the Canal. In this case, "the west" from which he returned referred to Portland, Maine. This common Maine coast term, usually expressed "the westard," applies to any relatively nearby place lying in a westerly direction. We must not lose sight of the fact that former Lubeckers sailed the seven seas, trekked to the California gold fields and by no means regarded such places as Boston or New York as the extreme limits of their travels. This shows perseverance and thrift. "Thursday the Tug boat *Sea King* brought one of the rafts which she had lost last week."

Other items include these. "Mr. Cushing has shifted the position of his saw mill and has it running once more." "A large amount of lumber has just arrived at the E.M.S. Co.'s wharf and is neatly piled, giving the point the appearance of a lumber yard." "Of those who come here in search of work each day, more arrive than depart. The newcomers find it difficult to let boarding places and a good one here would be a paying investment." "The following Lubec people are employed at the Canal: Calvin Trecartin, boss carpenter, Bert Knowles, Frank Trecartin, and Will Hunt, carpenters, F.N. Gillise, bookkeeper, Edward Mulholland, blacksmith, and Willis Armstrong, engineer's helper."

This last August 1948, two women visited the site of Plant No 2. One of them, whose maiden name was Swanson, lived in Cromwell, Connecticut, as a child. Her family was one of those that came here

200 Years of Lubec History, 1776–1976. Courtesy of the Lubec Historical Society.

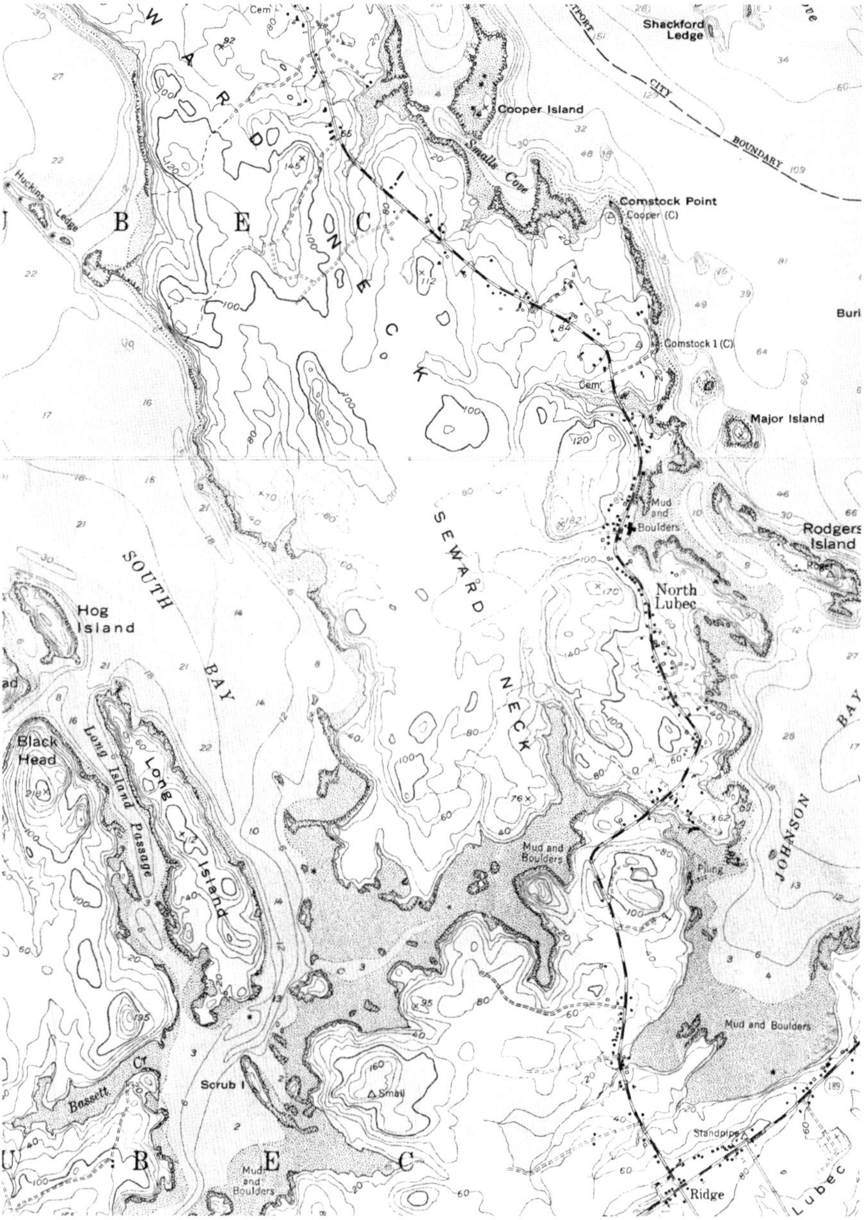

This 1949 U.S. Geological Survey map shows sites of the complete Klondike Plant No. 1 and the second incomplete Plant No. 2, the latter at the Canal. Note how the waterway cuts across Seward Neck, connecting Johnson's Bay on the east with South Bay on the west. North of Plant No. 1 at Comstock Point is the site of the old North Lubec ferry wharf. Indicated south of Plant No. 1, at the inlet close to North Lubec Road, lies the large Lawrence Brothers sardine canning plant.

to work on the project, and she was interested in locating any remains of the house they occupied. She could remember that there was a sort of basement in which many boxes were made. In company with Mr. M.P. Lawrence, they called on Mr. Asa Ogilivie to talk over those former years. Shortly after, the following item came to notice, which should help to clear up the matter of location.

"Mr. Swanson of Cromwell, Conn. is to rent the rooms over the carpenter shop. 32 men will be accommodated and the house will be ready the last of the week."

THE *GOLD BUG*

Article 13, March 10, 1949

"The E.M.S. Co.'s launch *Gold Bug* was given its trial trip last Tuesday by Mr. Frank Hallet, her builder." So reads an item, spelling and all, from the *Lubec Herald* of June 28, 1898. For many years thereafter, the motor launch *Gold Bug* was a familiar and attractive sight sailing about the bay. The *Gold Bug* was the first power launch built in this region.

According to information received from Mr. Edward Hallet, she was built by his father, Captain Frank Hallet, by order of the E.M.S. Co. on the recommendation of Mr. A.N. Pierson. The building took place in the spring, probably requiring two or three months. Most of the lumber undoubtedly came from Portland. The framework was of white oak with hard pine planking.

The *Gold Bug* was about 38 feet overall with a nine-foot beam and a draught of about 3 and ½ feet. She was equipped with a ten horsepower 4-cycle Globe gasoline engine. This boat was housed over two-thirds of the way and had an open cockpit aft.

Hotel Nemattano, Lubec, 1891. The New England YMCA built this timber hotel, also known as Ne-Mat-Ta-No, on North Lubec in 1890 with an 1891 season of July 11 to September 1, later July 1 to October 1. According the book *200 Years of Lubec History* (Lubec Historical Society, 1976), the rooms were $2.00 and $2.50 a day or $9.00 to $12.00 a week, including open fireplaces, baths, a dance pavilion, a bowling alley, a pool, tennis and deep-sea fishing. The five-hundred-acre facility burned on Saturday, August 6, 1921, then unoccupied. According to the August 11 edition of the *Lubec Herald*, "The big hotel went in a very short time, a spark from the fires setting fire to the roof." Fires were common in the region that summer, the driest since 1894. *Courtesy of the Lubec Historical Society.*

The cabin was about fifteen feet long. The engine room was between this and the cockpit. Probably the selling price was $2,000 to $3,000.

Captain Hallet operated the *Gold Bug* for the E.M.S. Co. for the few short weeks before the company suspended activities. The launch was used for running about on errands and for towing. For about a year after that, Captain Hallet had her for his own use. Among other things, he carried mail between Eastport and Lubec and ran excursions for the Ne-mat-ta-no Hotel.

The Globe Canning Company then acquired the *Gold Bug*, and she was run for a time by Mr. Kilby Coggins as captain. Following that, Captain Norman Lank of Campobello took over. The company had the boat remodeled, after a time, and renamed the *Globe*. She was used to carry the businessmen from Eastport to the factory and to perform other services for the company.

The Globe Canning Company sold the *Gold Bug*, or rather, the *Globe*, to a St. John, New Brunswick concern, and when last reported she was used for towing logs on the river.

What multifarious works this E.M.S. Co. performed. The bridge, the telephone and causing the *Gold Bug* to be built. Who actually gave her that name? How sad, in some ways, that this extraordinary company did not last longer.

The Dummy Dynamo

Article 58, January 19, 1950

It seems that there was a dummy dynamo at Plant No. 1 that was supposed to furnish the power for the accumulators that lay beneath the buildings and thus remove and deposit the gold content of the salt water. Whether it was really connected with the underlying apparatus or was arranged as something to show to interested visitors you may judge from the ensuing account:

"Officers of the Company Deny That There Ever Was Such a Thing. Stockholders Take Hope."

The story of the "dummy dynamo" from Lubec is branded as an idle dream by the officers of the company.

"H.H. Jones, who was alleged in the story to be the electrician who worked the "dummy apparatus," and who has recently departed from North Lubec, is said by the officers of the company to have had absolutely nothing to do with connecting this apparatus, as described in the story.

"The dynamo was put in by a Portland electric company. Sometime later, when a man was wanted to go to Lubec to do some general work, the officers of the company at the home office in the Exchange Building (Boston) sent word down to a Boston firm that they wanted a man to go to Lubec to do work. R.H. Jones, who had formerly done some work at the Quincy House, and who had been in the employ of the electrical firm about four days, was sent to Lubec. He did the work and finished up in a few days.

"After he finished, he asked the manager, Mr. Pierson, Mr. Ryan said, for a steady job with the company. A man was wanted for a post of night engineer and man of all work, and Jones was engaged. He did his general work at night, but he did not have to do with connecting up the dynamo or operating it.

"The officers of the company said last evening that the dynamo certainly did furnish power and generate electricity. It was the only source of the current which kept about 20 lights burning brightly every night from the opening till the closing of the plant a few days ago."

At least, the dynamo did something. This article appeared in the *Boston Herald* of August 4, 1889. Whatever went on, it was the part of wisdom to have the employees think that all was well. Too many in on the real nature of the doings would have injured the company's standing at an earlier date. Without exception, the employees said that they did their work in good faith.

Big Jack Ives at Plant No. 1

Article 59, January 26, 1950

In trying to analyze the affairs of the E.M.S. Co. after the collapse, Mr. Arthur B. Ryan, the president, said, in part: "The night watchman of the plant is, apparently, an honest, straight forward man, who has never given us the slightest reason to believe that he was in any manner connected with Jernegan and Fisher in the scheme. Yet it was at the same time apparent that the gang was bound to try to implicate him by referring to a person in their letters who might be accidentally taken for him.

"When the plant was first opened, Fisher insisted upon importing this watchman from England for the job, because he said he could trust him, and there would be no danger from him, because he was a perfect stranger to the scheme. This man was known as John Ives, whom he was told and believes, was formerly in the English army, serving the last of about nine years in India.

"Fisher said 'My friend from India' would receive a furlough until September, and his time would run out a day or so later, so that he would not be obliged to return to do any further service in the Army.

"Ives came over, and he, like the English bookkeeper, Firmer, who left the employ of the company three months ago and returned to England, served faithfully and well. No fault can be found with them, while the letters of Jernegan and Fisher to Phelan would seem to indicate that they were trying to throw blame upon Ives and Firmer to divert suspicion from themselves."

Italian Laborers

Article 15, March 24, 1949

During those spring and summer months of 1898 when Plant No. 2 was being constructed for the Electrolytic Marine Salts Company, many laborers were employed. Of some 700 men, approximately 200 were Italian immigrants.

This was shortly after the time when the Italian immigration was at its height and before the time when organized efforts were made to have them attend "Americanization" classes. Hence, most of them spoke English brokenly, if at all, and their ways and appearance being different, they were often regarded with suspicion.

They did not all appear at one time but came by groups at intervals. There were two large camps, or camping areas, where they stayed, one back of Jimmy Dodge's and one on the hill back of Susie Avery's home.

Again quoting from papers of that summer, we glean what few glimpses we may. The June 7 *Herald* informs us that "a laughable sight was witnessed one day last week in the Canal. A party of Italians were hauling on a rope (Oh, *Herald*, ye also made errors in grammar) fastened to one end of the scoops, when the rope parted and the entire crew fell over into about a foot of muddy water. They were a sorry sight when they got straightened out." Seems as if people laughed immoderately at minor misfortunes and practical jokes then. Now, they would probably summon the Red Cross, and the victims would doubtless sue the company and everyone would weep.

Almost every week the workmen were mentioned. This is found on June 14. "It has been remarked more than once that there is less fighting and quarreling among the three hundred men employed there than could be found among the same number of men in any other place. Even the Italians do not have many "rows" and so far have given the people of North Lubec no cause to complain of them." This goes to show that hearsay and prejudice played a large part in the lives of most people. In recent years, vastly different views are taken in regard to people who come as strangers to this country.

"The employees at the Canal are now working under the shadow of Old Glory. The flag was brought from

Portland by Mr. Shanahan on his return from that city (June 28)." The E.M.S. Co. certainly had all the trimmings.

On July 5, a *Herald* item read, "There are now employed here more than five hundred men, and new men are being hired every day. Nearly all the employees are Americans but a number of Italians are working here. Last Thursday a gang of thirty sons of Fair Italy arrived from Machias where they had been at work on the railroad but preferred the better pay given by the E.M.S. Co., so migrated to the Canal." No oral stories have been handed down to indicate that these people ever caused any particular trouble.

An interesting item from an Eastport column in the *Bangor News* for May 17 observed that "one of the most closely guarded articles of personal property and a noticeable sight on the arrival of the different crews of Italians to this city during the year is their umbrella, a cheap looking and faded out affair on the average, but valuable to the dark skinned laborers, who carry it always in their tramps to and from this city. Frequently an Italian workman comes down for a brief visit on foot from the contractors camp near, but rain or shine, the cotton umbrella is rarely forgotten, if one is owned."

The next write-up is taken from Eastport items in a daily paper after the cessation of activities at Plant No 2. "A large number of rough looking Italian workmen were in the city all day Wednesday and left later for the West with their baggage." The laborers had been at work for several weeks past on the construction on Plant No. 2 at North Lubec for the Electrolytic Marine Salts Company which went up in the exposure of last week.

"A report reached this city that one of the Italians had been murdered by his companions, but a careful inquiry had been made about the works…without finding out anything further, yet the story was widely circulated here Wednesday. A fight did take place there early in the week between Italian workmen but aside from a discolored eye, bleeding nose and four front teeth lost in the fracas, no other injuries were sustained by the dark-skinned contestants. It is generally understood that thieving was quite extensively practiced by many of the Italians before they left North Lubec as one workman can vouch for. He was asleep Tuesday afternoon on a bank near the new plant and when he awoke was minus his shoes, which had been unlaced and stolen. Hats and clothing were also articles of seizure frequently, and when several cases of theft became known it is to be wondered at that violence was not offered. The crew of more than six hundred laborers contained many tough characters from the west and how the work was conducted so successfully without trouble is to be wondered at."

Quite an item, and what things they printed fifty years ago. (This does not sound much like Brotherhood week.) Probably this was…a case of "the first one up-the best dressed." A little good natured playfulness among themselves. In all, those laborers of a half century ago lent picturesqueness to this already colorful project.

CHAPTER 4

Prescott, Prospectus and Plant No. 2

Plans for Plant No. 2

Article 11, February 24, 1949

How much speculation and suppressed excitement there was in Lubec when it was learned that the second and larger Plant No. 2 would actually be located at the Canal! Insofar as possible, the exact comments from the newspapers of the time will be quoted so that the picture may appear as it really was. Again, referring to the *Herald* of February 15, 1898, theorizing ran thus: "In speaking of the proposed plant at the Canal, officials say that if erected, it will contain 5,000 machines with a capacity of one million tons of water. The construction of such a plant would give employment to hundreds of men, and leave in our town large sums of money. It is safe to say that $4,000 would find its way each week into the pockets of the laborers, should the proposed plant be built."

There must have been many private plans as to the uses that would be made of the money so providentially sent. Work began promptly. The papers of March 1 had this to say. "Now that the commercial success of this process has been clearly demonstrated, the Electrolytic Marine Salts Company is perfecting plans for the erection of a mammoth plant down at the Canal, on what is known as the Fowler property, the site of the old plaster mill. [Early settlers prepared plaster from natural limestone in the area, leading to commercial processing when tidal power was harnessed. When the limestone ran out, the mill imported raw material from as far away as South Carolina.] Here is to be erected a plant with a capacity of 10,000 machines and capable of extracting $10,000 worth of gold every twenty-four hours. A crew of 500 men will be put on the first of April and it will take the remainder of the year to complete the work, entailing an output of over a million dollars."

According to an item of March 29, "Mr. A.B. Ryan, Pres. of the famous E.M.S. Co., accompanied by Mr. Sawyer, was in town Tuesday and Wednesday of last week. Work on a large scale will begin the first of next month (April). Mr. Jernegan is still in Boston and will not return for several weeks." Undoubtedly, he was busy getting more gold there than he ever was able to lure from the sea.

All our neighbors were interested. Frequently, the *Calais Times* made note of the doings of the E.M.S. Co. Here are samples of items written in March and April: "A.N. Pierson, who is financially interested in the Lubec Plant for extracting gold from sea water will pass the summer here (North Lubec) with his family, and has arranged to close his residence in Cromwell, Conn. Mr. Pierson was in this city (Calais) last week.

"D.W. Clark inspected the plant of the Electro-Marine (that's the way they spelled it) Salt Co., at North Lubec, Friday, and personally handled some of the gold that had been extracted from old ocean.

Site of Plant No. 2, looking east into Johnson Bay off Cobscook Bay. Reproduced in the 1976 book, *200 Years of Lubec History, 1776–1976*. *Courtesy of the Lubec Historical Society.*

While he was not permitted to see the process, nevertheless he was convinced that the company had a bonanza. He said there was no doubt they are getting the gold, and are daily adding new machines and thus increasing profits. Mr. Clark says 2,000 machines will be in operation this summer…"

Well, that is only one more proof that "there is one born every minute," as P.T. Barnum put it.

Work went on, and an Eastport column fills in part of the story, so much of which is forever lost. A reporter observed that "steam tug *Sea King* arrived from St. John [New Brunswick] Monday with 150,000 feet of logs and 400 pieces of spiles for the E.M.S. Co., the new Klondike at North Lubec, near this city. The lumber is from Allston Cushing, Esq. who has a contract to furnish 8,000,000 feet to be used in building the new factory for this company. The building will be 120 x 30 feet and is now being erected by a large crew of carpenters. Mr. Cushing of St. John has a Monarch 60 horsepower boiler and a 50 horsepower engine erected on the ground near where his portable mill is sawing up the logs into suitable lumber. About 35,000 feet can be turned out each day and large rafts of logs will be towed down the Bay of Fundy to North Lubec, near the new Klondike, during the month to be sawed up and used in building their new plant. Just at present, North Lubec, near the new Klondike, is a very busy place and plenty of work is had by the many laborers there."

Thank you, Eastport. Such goings-on. As the weeks sped by, we read of boarders, housing projects, a barbershop, a millinery shop, tugging, towing, drilling and blasting and the coming of foreign laborers.

Readers of Boston papers noticed the following obituary on February 20: "Edgartown, Mass. Dr. Marcus Wilson Jernegan, 76, professor emeritus of the University of Chicago and a specialist in American Colonial history, died here last night at his family estate." (Marcus W. Jernegan was a brother of Prescott Ford Jernegan, and he will figure in our story of the E.M.S. Co. later on.)

"An authority on various leaders of the Colonial period, he unearthed evidence that proved Franklin's discovery of electricity had been accomplished before he knew of similar French efforts. A native of this town, he was graduated from Brown University in 1896 and received his doctorate from the University of Chicago in 1906. Two years later he returned to Chicago where he remained as a history professor until his retirement in 1937.

"He was a member of the Illinois State Council of Defense, American Historical Society, Phi Beta Kappa, Colonial Society of Massachusetts, and the Quadrangle Club. Dr. Jernegan was the author of several books on history, including a *History of the Tammany Societies of Rhode Island*, *Laboring and Dependent Classes in Colonial America*, and *The Growth of the American People*.

"He leaves two daughters, Mrs. Bruce W. Cook of East Longmeadow and Mrs. Edward Krueger of Chicago."

The Canal Bridge

Article 8, February 3, 1949

The Electrolytic Marine Salts Company was almost always styled the E.M.S. Co. by the *Lubec Herald* in 1898, and from now on, for the most part, this review will do the same.

In the early part of February 1898, "one of the editors of the *Lubec Herald* took the 1:15 boat to North Lubec to see the work that is being done… at the 'New Klondike.'" While in conversation with one of the officers of the E.M.S. Co., he was told that although the company had purchased the water privilege at the Canal, they might conclude to locate elsewhere.

"For instance," the company's representative said, "the condition of the wooden bridge, which should be replaced with an iron one, will influence our decision. We have a plan to help the Town replace this bridge with an iron one farther up the Canal. The Town might pay half, and our company half, making our share a present to the people of Lubec. But this is but a possibility, as we have by no means decided to locate there."

As a result of this visit, the following sage council appeared in the editorial column of the February 15, 1898 issue of the *Herald*. "A liberal policy is always the wisest one in dealing with capitalists who contemplate building up an industry of the importance of the one being established

Ferry, North Lubec, circa 1895. The Passamaquoddy Ferry Company, serving Lubec, North Lubec, Eastport and Canada's Campobello Island, began operation in 1882 under sails. Eventually, the first steamer *Spray* gave way, in 1891, to the fine new *Lubec*. Built by Joseph Dyer of Portland expressly for this route, it was sixty-nine and a half feet long, eighteen feet beam and fifty-one tons with fifty horsepower and carried up to 350 passengers with the comforts of steam heat and electric lights. *Courtesy of the Lubec Historical Society.*

at North Lubec. We, as a town, should heartily cooperate with the gentlemen comprising the Electrolytic Marine Salts Company, in building the proposed iron bridge at the 'Canal,' North Lubec. It is understood that this company will bear half the expense of a bridge costing $3,000. This is a most liberal offer and should be at once accepted if presented at our next town meeting."

It appears that some of the town fathers had feared that the construction of dams or gates to hold the water in the Canal would affect the wooden bridge, and it likewise appears that the "Klondike" said that they would build one that the dams wouldn't affect. We can do no better than to trace the story of the building of the Canal Bridge

This mystery photograph bears the pencil annotation on the rear: "Built at time of gold from seawater swindle." While it appears to be a new bridge across the Canal, clearly it is not the iron structure underwritten by the Electrolytic Marine Salts Company. Nor do the geological features resemble those at Mill Creek, site of Plant No. 1. North Lubec is today far more timbered. *Courtesy of Davis Pike.*

through the items that follow. These were all taken from the *Lubec Herald*, and most of them were under that inimitable column, "Nuggets from the Canal."

April 5, 1898: "Mr. Town, representing the Berlin Iron Bridge Co., was in the village Saturday, and in company with C.H. Clark, Esq., one of the committee appointed to look over the plans for the proposed iron bridge at North Lubec, went up to the Canal and carefully looked over the location. Mr. B.B. Reynolds, one of the committee, informs us that he believes the bridge can be built for $3,000. [This company began manufacturing bridges at its plant in East Berlin, Connecticut, not unexpectedly only a few miles from Jernegan's church in Middletown, in 1885. The firm manufactured over one thousand such highway and railroad bridges before it was absorbed by the American Bridge Company in 1900. Former employees than started a new firm, still in existence under the name Berlin Steel.]

May 24, 1898: "The action taken by the last special town meeting regarding the proposed steel bridge at the Canal, North Lubec, will be considered. Another special meeting will be held this afternoon and the liberal offer made by the Electrolytic Marine Salts Company will be met in as liberal a spirit. There is no doubt that the town will vote to bear one half the expense of a new bridge. We believe that time will prove these gentlemen of the E.M.S. Co. to be staunch friends and people who will take a warm interest in the welfare of the Town.

June 7, 1898: "Work will begin on the new steel bridge at the Canal in a short time. This will be one of the best constructed and most durable bridges in this section of the state."

June 21, 1898: "Work on the new steel bridge is well under way. It is expected that the bridge will be completed before July 4th."

June 28, 1898: "Mr. Merrill, Engineer-in-charge at Plant No. 2, chartered the boat *Julius Wolff* and made a trip to Red Beach last week to purchase some of the Red Granite Co.'s building stone for the new bridge."

Evidently, the bridge was not finished by the fourth. The next item is taken from the *Herald* of July 5, 1898. "The runaway in which Miss Mame Bennet [Mame Seeley Bennet (1879–?), daughter of legendary Lubec physician Eben Bennet] and Miss Fannie Brown were involved, occurred here at the bridge…(It must have been the old one)…Tuesday evening of last week. The ladies were driving over the structure which has been narrowed so that but one team can pass at a time. Directly at the end of the bridge a large crowd was assembled, and this the girls were trying to avoid when they collided with an approaching team. Both young ladies were thrown to the ground and the horse started to run away but was caught and

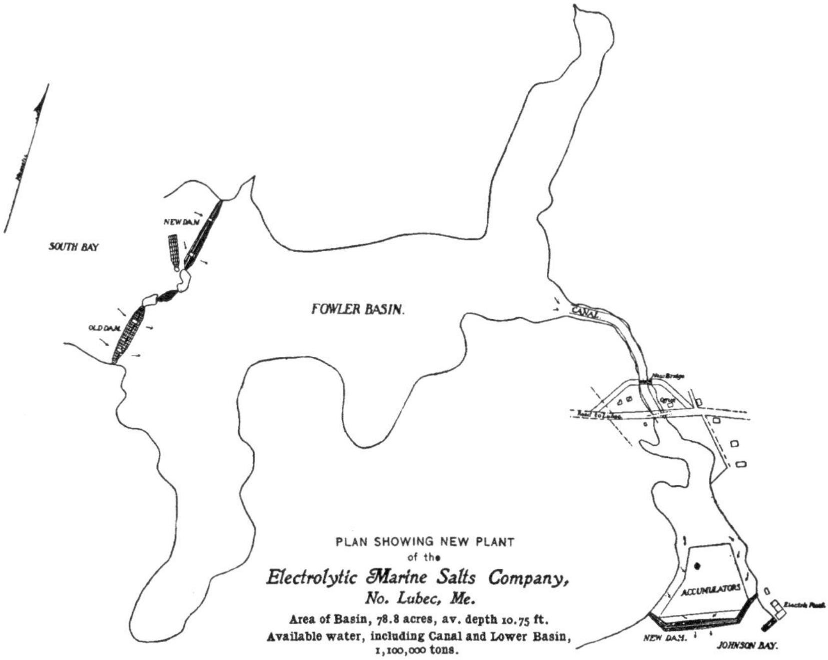

Map from the prospectus of the never-completed Klondike Plant No. 2. In 2013, North Lubec Road runs straight, as shown on this 1898 map. No evidence remains of the new road and iron bridge. At the point to the left of the Canal where this road rejoins the main road is Plaster Mill Road, still extant. The partially dotted way opposite is now known as Klondike Road. A house now stands where "Office" is indicated, 298 North Lubec, but was apparently built later, as a concrete block foundation is evident. *Courtesy of Edith Comstock.*

returned before going far. We understand the girls were not injured, but were badly shaken up."

It was more than two weeks after the debacle of the E.M.S. Co. that the final bridge item appeared on August 16, 1898: "The new steel bridge at the Canal, and which will remain for many years a monument to the E.M.S. Co., was completed last week. The bridge is a beauty, and will last a lifetime."

Well, the bridge did stay in use some forty years, and we are grateful to the E.M.S. Co. It was then replaced with the present one with the filled-in foundation so that boats can no longer sail up the

Canal without being hoisted across the road. The old iron bridge may have looked a little fairy like and fragile for today's vehicles, but it held up perfectly well, and no car ever crashed through its fancy black iron rails.

A Few Social Activities

Article 18, April 14, 1949

During the spring and summer of 1898, there were many visitors from "away." Certain of the officials of the E.M.S. Co. spent no small amount of time entertaining them and showing them such parts of the gold plants as they wished them to see. Courtesy and dignity, in keeping with so famed an industry, were matters of course.

Various company officials came and went. The "Boston Boat" profited by reason of the E.M.S. Co. "In addition to work and conducted tours, time was found for diversions such as outings and horseback riding. Whether they gave any sign or not, the local inhabitants must have felt a news gatherer's interest in the activities.

Although other papers have accounts of more of the visitors to North Lubec that summer, we shall, for the present, confine ourselves to the fragmentary items of the *Lubec Heralds* at hand. "Mr. A.B. Ryan, Pres. of the famous E.M.S. Co., accompanied by Mr. Sawyer was in town Tuesday and Wednesday of last week. Work on a large scale will begin the first of next month. Mr. Jernegan is still in Boston and will not return for several weeks." (March 29, 1898). Mr. Sawyer was one of the directors and the secretary of the company. Mr. Jernegan came and went often, it is noticed, and no one paid any special attention to this as being unusual. He was noted for announcing his times of departure and his proposed trips in advance, even the final one.

In the *Herald* of April 5, 1898, we find that "Mr. E.E. Trecartin, of the firm of Trecartin Bros., has been at work plumbing the Bangs house recently purchased by the E.M.S. Co. at North Lubec. [For decades, Edward Trecartin (1865-1949), with his brother John (1869 or '70-?), ran Trecartin Brothers Hardware in the building now at 32 Water Street.] This house, when completed, will be occupied by Mr. Pierson and family who will move from Cromwell Connecticut to North Lubec early the coming season."

The roving Mr. Jernegan is again noted in the issue of June 21: "P.F. Jernegan, wife and child, arrived from Boston by Thursday's boat." It may be wondered how many trips he made between March and June.

Our neighbors were interested, too, in getting first-hand information, and doubtless remembered that one picture is worth ten thousand words. "Mr. Rumery, junior partner in the firm of Rumery Bros., Eastport, was in town last Wednesday looking over Plant No. 2. [An Eastport 1901-2 business directory lists the Rumery Bros. clothing firm at 52 Water Street.] Rev. J.A. Ford with a camera of generous proportions was also viewing the work going on at the Canal last Thursday."

Officers and stockholders made repeated trips; it seems that "visitors from the west have been numerous here the past week. A large party among whom were Mr. Sawyer, Mr. Brockway, and a number of other gentlemen, with half a dozen ladies, arrived by Tuesday's boat and remained until Friday." Where did they stay? Maybe at your house, or at mine. Do we ever notice the word "gentlemen" in items of the present? The item sounds funny; undoubtedly those ladies were the gentlemen's wives.

In the same column, it states that "nearly all the visitors who were in Mr. Sawyer's party returned to Boston by Friday's steamer." Think of the boat service in those good old days and weep. [Lubec never had railroad service. The Eastern Steamship Company provided regular service from Boston to St. John, New Brunswick, with regular stops at Lubec from 1893 until 1931.]

A July 5 copy notes that, "Mr. A.B. Ryan, Pres. of the E.M.S. Co. returned to Boston Friday." Mr. Ryan's family also spent that summer at North Lubec. There were three small boys. One of them, Arthur B. Ryan Jr., in a recent conference, recalled being much intrigued by the sardine factory. Mr. Ryan said that their family lived in the Reynolds house that summer. Which Reynolds house, he did not know. It may have been the one that was built by B.B. Reynolds at that period, or it may not.

Not all of the important people who visited the Canal came from Southern New England. The *Herald* put it this way: "We had some distinguished company here Wednesday of last week, the party being composed of Messrs. Julius Wolff, B.O. Bowers, C.J. Staples, and B.M. Pike. [The ferry service connected the Cobscook Bay communities of Lubec, North Lubec and Eastport and the Canadian Campobello Island.] In 1882, the Passamaquoddy Ferry Company was formed by Julius Wolff, a pioneer of the sardine industry, and Bion M. Pike, long time sardine mogul and businessman of Lubec who later became sole owner of the company." [The first sardine factory, the Lubec Packing Company, was started by Julius Wolff with Moses P. Lawrence and Henry Dodge in 1880. It was located at the "Acre," also known as the "Devil's Half Acre," at the north side of the inlet to the Canal, beyond the end of what today is known as Klondike Road.]

Who remembers the boat known as the *Julius Wolff*? In regard to the famed July holiday it is recorded that "R.D. Shanahan, George A. Merrill, and a number of other gentlemen chartered the *Julius Wolff* and with her went to Calais to celebrate the glorious Fourth." They must indeed have enjoyed the trip up the St. Croix. [The seventy-one-mile St. Croix River forms the easternmost section international border between Maine and New Brunswick.]

BAIT FOR "SEA GOLD FISH"

Article 54, December 22, 1949

"MEANS BY WHICH IT WAS SOUGHT TO EXTRACT MONEY FROM POCKETBOOKS OF THE CREDULOUS"

From the *Boston Herald* of August 1, 1898, is an article leading up to and quoting the prospectus of the Electrolytic Marine Salts Company written in a flight of fancy by Prescott Ford Jernegan himself. "Never did fishermen bait his traps," states the *Herald*, "with more alluring or attractive morsels than did the Rev. P.F. Jernegan tickle the fancy and stimulate the greed of victims with his brilliant and enticing prospectus of the Electrolytic Marine Salts Company, now a practically defunct organization, with the reverend promoter flown to foreign parts, an alleged swindler of the first magnitude."

A glance at the document, which was the only too powerful means of extracting gold from pocketbooks, not seawater, is pathetically interesting at this stage of the proceedings. Therein was furnished an exhaustive treatise on gold in general, gold in seawater in particular, a scientific sketch of the proposed process for removing the precious metal from the briny solution on a paying basis and, running through it all, the unanswerable story of personable experiment and personal experience.

The bait for the "seawater gold fish" was labeled. It read, "A sketch of the Discovery of a Commercially Profitable Process for the Extraction of Gold and Silver from Sea Water." The author

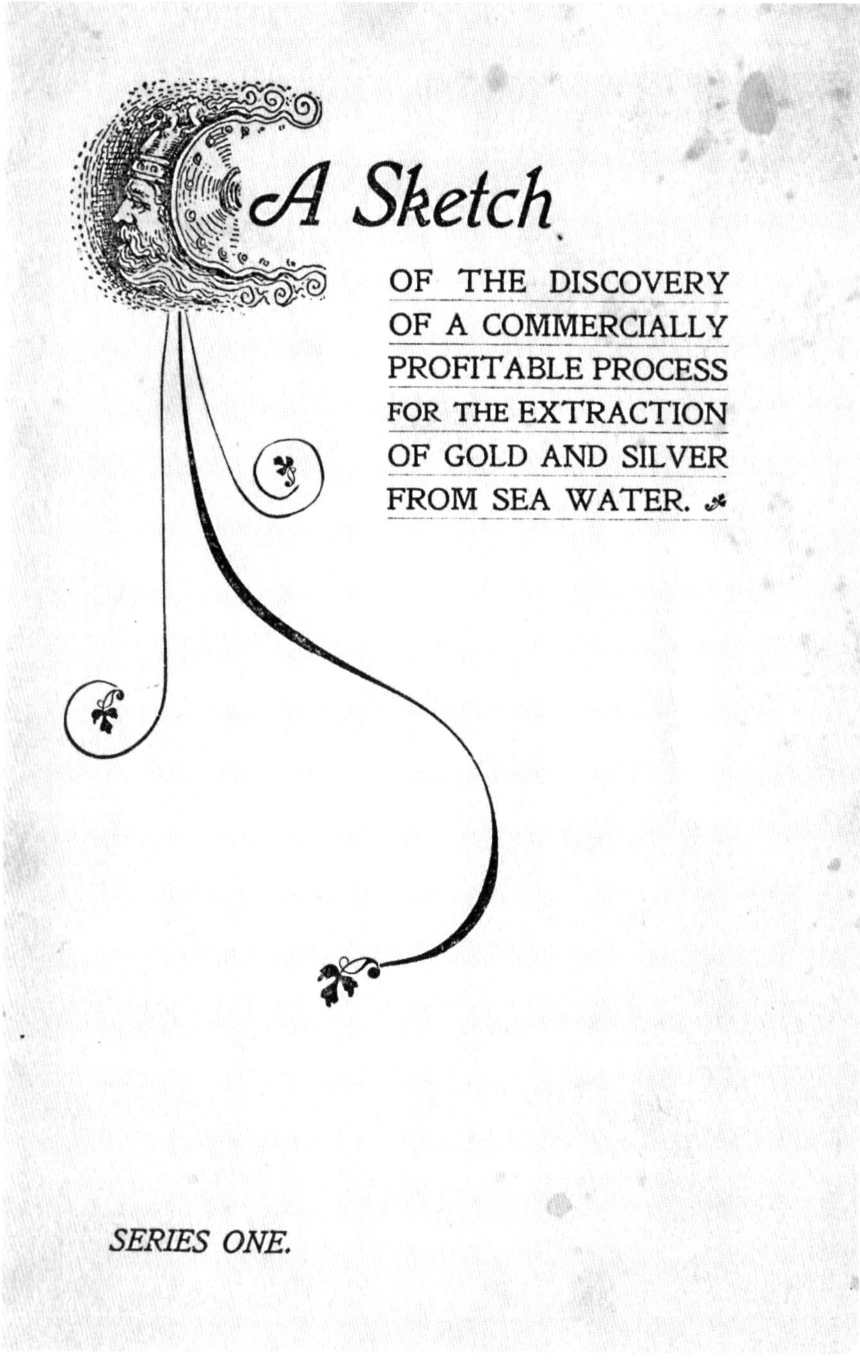

A Sketch

OF THE DISCOVERY OF A COMMERCIALLY PROFITABLE PROCESS FOR THE EXTRACTION OF GOLD AND SILVER FROM SEA WATER.

SERIES ONE.

Cover of the prospectus for Klondike Plant No. 2. *Courtesy of Edith Comstock.*

of the ingenious work even boasts a preface to his tale. He says, "In the summer of 1896, while recovering from a[n] attack of typhoid fever, the attention of the writer was called to the existence of gold in sea water. A taste for science and familiarity with the sea led to an instant appreciation of the vast commercial possibilities of this fact.

"From that moment researches have been conducted with the utmost energy and at the expenditure of many thousands of dollars, with the result that profitable methods of extracting the precious metals from the ocean have been discovered, and the Electrolytic Marine Salts Company has been incorporated to develop these discoveries."

"He then started in with the usual array of 'firstlys,' 'secondlys,' 'thirdlys' and 'finallys,' common to the old time preachers, winding up with the customary resume of things already oversaid. The 'firstly' in this instance was a contrast of the difficulty of ordinary gold mining from the rock, compared with the marvelous ease with which the author promised to exhort precious tribute from Neptune.

"In the 'secondly' part of the shrewd divine's pretty production the initial statement was that 'the waters of the ocean contain gold.' This fact was established by the quotations from three well known encyclopedias, which follow, and a quoted article by Professor Henry Wurtz [American inorganic chemist, 1828-1910] in the *Engineering and Mining Journal*, and extended reference to the investigations of Mr. Edward Sonstadt [British chemist who discovered, in 1872, gold to the extent of one grain, a minuscule unit of weight, per ton of seawater] and of Prof. A. Liversidge [Archibald Liversidge, 1847-1927, chemist and mineralogist] of Australia.

"Having established the premise that 'there is gold in sea water,' without the peradventure of a doubt, our friend of the Jules Verne tendencies broke the fetters which bound him to the realm of fact and soared majestically upward to the enchanting, but cloudy, regions of fancy."

Next week's *Lubec Herald* will carry the story of the prospectus. (That is what typhoid fever can do for a person.)

JERNEGAN'S PROSPECTUS

Articles 55 and 56, December 29, 1949, and January 5, 1950

Wouldn't it have been interesting to watch Mr. Jernegan write his prospectus? What must his facial expressions have been? Do you suppose he and Mr. Fisher ever indulged in some good private laughs as they saw what developed from their ideas? Did Jernegan think along the lines of extracting the existing gold so much that he had faith that he exhibited to the stockholder? Maybe he felt that he could lay the failure partly to the imperfections of Fisher's apparatus.

At any rate, he made some earnest advances and preparation and prepared his treatise, which follows:

"A Task for the Imagination.

"One is at a loss to comprehend the enormous wealth floating in solution in the ocean. At the lowest estimate, a cubic mile of sea water contains gold to the value of $65,000,000. It is probably nearer the mark to place it at $100,000,000.

"There is enough gold in the waters of Long Island Sound to pay off the national debt and leave a larger gold reserve in the treasury than the government has yet possessed.

"The waters that sweep in and out of New York Bay daily contain enough gold to buy all the ships and merchandise borne on their surface.

"Massachusetts Bay holds enough of the precious metal to buy all the real and personal property in the entire state. Acre for acre, the waters of the Bay are worth more than the land of the state.

"A Practical Undertaking.

"In reference to personal investigations, the writer needs only to remind the reader that anything like an explicit revelation of the methods employed would be a betrayal of the interests of the stockholders of the company which has acquired the use of these processes. He desires, however, to direct especial attention to one principle that has guided all the investigations, viz., that no theory was to be accepted until it had been tried on a practical scale at the seashore under conditions that would conform to those of a working plant.

"With this principle in mind, practical tests have been made in Narragansett Bay, Long Island Sound, Passamaquoddy Bay, Whitable-by-the-Sea, England, and Sea View on the Isle of Wight, as well as several other places. As a result of these experiments, hundreds of dollars worth of gold have been extracted at an expense that leaves no doubt of the practicability and profitableness of the processes.

"The writer has been assisted in these experiments by a co-worker and indirectly by prominent chemists in England, where six months were spent in perfecting where advantage could be taken of the latest and most complete chemical laboratories."

"The process which shall be employed in the immediate future uses a series of identical pieces of apparatus, each of which is a complete machine in itself, and uses water independently of all the others. One of these sets of the full size that will be used in the largest plant that need ever to be built, has been tested at the very spot in Passamaquoddy Bay where our plant is in the process of installment. The results were eminently satisfactory. Before Jan. 1, 1898, it is confidently expected that a plant will be in operation which will extract a minimum of $100 worth of gold a day, at a cost that is a small fraction of the production. By the spring or summer of 1898 it is believed that a plant [Plant No. 2 at the Canal], the location for which is already acquired, will be in operation, producing $1,000 a day.

"The writer was in a position, nearly a year ago, to substantiate his claim for the discovery of gold from sea water. In the time that has since elapsed, he has expended thousands of dollars in order to establish a process that should be uniform in the rate of production, ample in operation, and capable of rapid expansion.

"A Credible Claim.

"To many the above claim may appear astounding, not to say incredible. Reflection will show, however, that it is not for a moment to be classed with the so called processes for transmuting the baser metals into gold. Attention is called to the following indisputable facts:

"1. The ocean does contain gold in uniform distribution and practically inexhaustible quantity.

"2. Any person can verify this fact at a small expense of time and money with the assistance of a chemist or assayer.

"3. Scientists have already discovered and published processes that applied on a large scale, very nearly make it profitable to extract gold from sea water. The problem is far less difficult than many that modern science is daily solving.

"4. The writer of this article has investigated this subject in a more practical way, viz., by making tests on a large scale at the sea shore than any other investigator, and at a much greater financial outlay.

"5. As a result, he has demonstrated to several gentlemen of ability and intelligence, who cannot possibly be suspected of collusion, the practicability of this process. These gentlemen have furnished their own chemicals and had the fullest liberty of access to all apparatus used, following the experiments with unremitting attention and testing the results through assayers of their own choice and the highest standing. In some cases, the discoverer has not been present at the test or known even where it was being held."

Undoubtedly there were other embellishments and conclusions to this intriguing exposition, but this gives us an idea. Most of the article is true except for the way that the gold got into the machines. We might wonder if Mr. Jernegan kept a straight face as he wrote this prospectus.

THE NEW OFFICE BUILDING AT THE CANAL

Article 14, March 17, 1949

The first office building of the E.M.S. Co. was at Plant No. 1 in the "Klondike" buildings themselves. Here it was that all the gold was analyzed, prepared into the $2000 bricks and sent to the assayer's office in New York.

When construction on Plant No. 2 went forward on such a grand scale at the Canal, one of the first projects was the new office building. It was located just south of what was the Canal Hall, and part of this "office" still stands. It was fully twice as long as it is at present and for many years was referred to as a tenement house.

In March 1898, a local item stated that "Mr. Calvin Trecartin, who has the contract to build the new office for the E.M.S. Co. at North Lubec, has now in his employ seven men, among them being Mr. Frank Trecartin. The building will contain four offices and be finished in hard pine."

Weeks passed until June 7, 1898, when the *Lubec Herald* was able to offer this description: "The new and commodious office at the Canal is fast nearing completion. The offices downstairs are finished, and the chambers above nearly so. This model office of its kind is situated just north of the bridge, and is painted a light grey with dark green finish. The offices are finished in hard pine and each is furnished with an abundance of desks etc. No expense has been spared to make the work pleasant for the employees. Entering the North door, the first door to the right bears the sign J.W. Gardiner, Surveyor of Lumber and Piling. Opposite is the office of Geo. A. Merrill, Engineer. Next is the contractor's office and across the hall from this is that of A.N. Pierson, Manager of Construction. A stairway leads to the roomy chambers above, which will be occupied by two of the clerks and families. At the end of the hall is a door opening into the general office."

That was a rather pretty building in its day. Many will remember it as being painted pink. Several families occupied "apartments" there. A few years ago the south end was torn down, due in part to a very shaky foundation. It is an honor, of sorts, to have lived in this office building of the E.M.S. Co.

To go back to 1898, on June 21, the *Herald* recorded that "Mrs. E.H. Bidwell son and daughter arrived from Connecticut last Thursday. Mrs. Bidwell is the wife of our genial timekeeper at Plant No. 2; they will occupy rooms in the upper part of the new office." At least, they had somewhat more than a month to stay before the abandonment of the project.

It may possibly be that the following family was another of the occupants of the office building, due to the character of Mr. Atwater's works. In that same issue we find that "Mr. C.B. Atwater, bookkeeper at plant No 2, is an amateur artist of no mean order. He recently took some excellent views of the Canal and its surroundings."

"Mrs. C.B. Atwater of Connecticut is expected soon to join her husband here. By the time Plant No. 2 is completed, we shall have a large colony of Connecticut people in Lubec." The very next week saw this fulfilled. "Mrs. C.B. Atwater and her mother, Mrs. Stebbins, arrived from Springfield, Mass. last Thursday."

It requires imagination to picture in excess of five hundred men being employed at the Canal fifty years ago; but in addition to building the gold plant itself, there was a housing project in progress as well as construction of the front and back dams. All of this had the affect of attracting commercial ventures also. It was a great era if a short one.

AN ARTICLE in the December 7, 1941 *Portland Sunday Telegram and Sunday Press Herald* interviews Richard M. Larrabee, eighty-seven, then a resident of South Portland. Mr. Larrabee worked on the project back in 1898 when he was about forty-four. Larrabee answered the reporter:

> *Did I see it? Why, I went down there and I worked on the place carpentering till it was closed up. They had about 800 men working and they put up all kinds of buildings and built a new steel bridge across the canal. They had a sawmill that sawed all the lumber. People would say, "Well now, there must be something in that or they never would layout all that money."*
>
> *There was a Baptist church in Lubec that was finished on the outside but never done on the inside and those men finished it all up. The head one said he was a Baptist minister and that's why they done the church…That was the man that run away with the money.*

Larrabee described the gold accumulators for the reporter:

> *They had little machines…in boxes about two feet square planted out in the salt water to draw the gold out of it as it sluiced through with the tides. Then they went to work somehow or other and put in little gold nuggets… People would go there to see it and they would take up a box and show them how much gold was made in 24 hours, and it made everybody crazy. Folks invested millions in it.*

Entrance to the Canal looking west toward Johnson Bay. The small promontory into the waterway, prominent on the maps, is the dark, heavily forested area on the right. Construction of Plant No. 2 began with the erection of a wooden dam across the wide entrance just beyond the promontory. *Photograph by Ronald Pesha.*

When they got good and ready Jernegan took all the money he could get hold of and skipped. Well, there was a great time in Lubec that night. About 2 o'clock fellers come around and told us workmen it had folded up. They paid off all the men but the investors lost every cent they put into it. Biggest swindling scheme we ever had in Maine.

Activities at the Canal

Articles 16 and 17, March 31 and April 7, 1949

The first part of the summer of 1898 was a time of "looking up" at the Canal. The E.M.S. Co. had numerous projects in the process of construction, and with hundreds of people employed at these activities, it is small wonder that enterprising folk, not connected with the company, thought of establishing business ventures for themselves.

There was need for much lumber for the considerable building planned in connection with the plant itself. The items below will help explain what was being done: "A large amount of lumber has just arrived at the E.M.S. Co.'s wharf, and it is neatly piled, giving the Point the appearance of a lumber yard" (June 21, 1898). "Mr. Cushing has shifted the position of his saw mill and has it running once more" (June 7, 1898). "Tug *Sea King* arrived from St. John with two large rafts for the saw mill" (June 7, 1898). This tug seemed to be on similar missions continually that summer.

"Messrs. McPhee and Gillise are having a boarding house built for the accommodation of the men employed in the saw mill. Mr. Angus Smith is doing the work" (June 28, 1898). Indeed, much of the lumber was used in what, today, would be termed housing projects. "The company has begun erecting a number of pretty cottages for their employees. They will build some twenty in all. They will be situated on the hill to the north of the site of Plant No. 2" (June 14, 1898). "The cellars for the new double cottages are being rapidly completed and work on the stone masonry will soon be commenced. These cottages are to be models of their kind, and very pretty… They are designed by John Johnson, boss carpenter at Plant No. 2. Mr. Johnson has been in charge of the carpenters since the company was organized, and built Plant No. 1 from plans of his own drafting" (June 21, 1898).

"Mr. William Aston, employed by the E.M.S. Co. as a carpenter, has returned to his home in New York on a business trip. He will be absent a month or more" (June 21, 1898). He must have returned about in time for the grand finale.

"Chas. Smith has charge of the work on the masonry for the new cottages and is pushing the work fast" (July 5, 1898). In this same edition, and perhaps in connection with the new office building, we read that "Mr. Ed Trecartin of the firm of Trecartin Bros. was at Plant No. 2 last week doing some plumbing for the company."

Coexistent with this building program was another at the same Canal area by a concern already established. "A number of camps have been built at the Half Acre, North Lubec, this spring, which would indicate that prosperous times are expected by the employees of C.L. Pike's factory" (June 7, 1898). "Schooner *Fred C. Holden* discharged coal at C.L. Pike's wharf last week" (June 21, 1898).

That term "Half Acre" is puzzling; most people have always heard of it as "The Acre." ["The Acre" is down Klondike Lane, on the east side of the North Lubec Road, just after crossing the Canal. Sardine factories were built at this site in the nineteenth century.]

The problem of providing food must have been a difficult one for many people, but we find an effort being made to solve it. "Mr. Geo. Boone, of Calais, is to build a meat market here shortly" (June 14, 1898).

In these items you have noticed the customs of the times of abbreviating the men's names when possible. Such a procedure would not be acceptable now. If there is one thing, more than another, that a paper of the present day tries to do, it is to get a person's name correct. That is not to say that errors do not occur, however.

Efforts at maintaining good personal appearances were not neglected. "A new barber shop is being built at the Canal" (June 21, 1898). Where do you suppose it was? "Mr. Harte has opened his barber shop and is doing a good business, judging from appearances" (July 5, 1898). Let us hope that Mr. Harte made a sizeable fortune before the fateful day of July 29.

The women were not neglected nor forgotten, for in other "Nuggets from the Canal" we see that "Mr. Winslow Morton is building a store on his property. It will be occupied by Miss A.M. Bucknam as a millinery establishment" (June 21, 1898). "Miss Annie Bucknam's new store is nearly completed and she will soon have an attractive line of millinery on sale." We wonder if she did? And if it was?

Cottages, the sawmill, the meat market and the millinery shop, all these were but sidelines to the main job of construction of the actual Plant No. 2 itself.

To prepare the foundation for the new building, or Plant No. 2 at the Canal, an underpinning of piling was needed. Beneath the building the boxes for the accumulators or gold-collecting machines would be placed. The work got under way. That many Lubec people found employment with the E.M.S. Co. may be gathered from various items.

"Mr. Jasper Myers has sold his pile driver to the E.M.S. Co." "Mr. David Tinker is in charge of the scow which will be used for driving small piles" (June 14, 1898). "Capt. Dan'l Johnson, Marshall Jenkins, C.A. Lamson and Lafayette Johnson are employed at the Canal" (June 21, 1898).

There certainly was much traffic by water, and many heavy cargoes arrived in addition to the logs and lumber previously noted. "Schooner *Annie*, with a cargo of brick, arrived at the E.M.S. Co.'s wharf last Tuesday morning" (June 21, 1898). "Sloop *Rocky Mountain* arrived here Thursday with a cargo of granite from Red Beach for the coping of the new steel bridge. [Red Beach, several miles south of Calais, Maine, is noted for its distinctly reddish stone.] "*Arthur C. Carlow* came from the same town to dress the stone. The concrete piers for the new bridge are rapidly nearing completion" (June 28, 1898). "The company are (maybe you would say "is") to have three large scows built, one of which is under construction, for freighting purposes" (July 5, 1898). "Tug *Sam'l B. Jones* arrived Saturday morning bringing the wrecked schooner *Henry*, with a hold full of piling for the E.M.S. Co. The *Henry* was owned by E.J. Sawyer of Jonesport, and was wrecked at Ragged Cove, Campobello, some three weeks ago" (July 5, 1898).

In addition to all the boating and building projects, the preparation of the Canal was in progress. To get the maximum results from the salt water, dams would be needed. Who has not heard of the Back Dam as well as the front one? The next item will give some idea of the magnitude of the work attempted.

"Mr. Wescott has charge of the crews who are drilling and blasting. 15,000 cubic yards of rock are to be removed from the Canal and five tons of dynamite have been ordered with which to do the work. Seventeen thousand cubic yards of earth will be removed, making 22,000 cubic yards of material to be taken from the cut. (Most mathematicians would arrive at 32,000 cubic yards.) Up to June 8, 1,764 cubic yards had been removed" (June 14, 1898).

"Otis C. Brown of Benton Falls, Maine, is foreman on the north half of No. 2 dam. He is well known throughout the state, having been engaged in building dams for many years."

Some of this work must have been dirty, and provisions evidently were made by this company that thought of everything. Make out of these items what you will, but not let anyone believe that no one in Lubec had ever had a bath before. "Klondike No. 2 has established free baths for its employees. These baths are much frequented and the bathers are always well applauded. Scarcely a day passes without somebody suddenly plunging into the water without giving any notice of his intentions to do so. The workmen are not the only ones who indulge in this diversion. The engineers have condescended to favor an admiring audience with a few graceful plunges.

Workers building Plant No. 2 at the Canal in 1898. The gentleman at right of center whose head and shoulders are visible under a light-colored hat is identified as Harry Meyers, father of Greta Wilcox. *Courtesy of Edith Comstock.*

The 1st Asst. Engineer showed his endurance one day by remaining in the water until pulled out by the impatient spectators" (June 21, 1898).

"Last Saturday, one of the crew on a punt suddenly slipped gracefully out of sight. On Sunday, the second engineer gave a striking example of sounding with a pole, which did not reach bottom, and which he followed, reappearing at length, and lighting a damp cigar to show that he was not supernatural. Bathing is becoming a fad and is now the proper thing, as is conceded by all who are in the swim." Here is a good chance to look up "bath" and "bathel" and its variants and to interpret these anecdotes correctly.

CHAPTER 5

Mischief, Maneuvering and Manipulation

Electrolytic Marine Salts Co. News

Article 19, April 21, 1949

All through the summer of 1898, there was unparalleled activity at the "Klondike" in connection with the new gold industry. Building and construction occupied many people. Associated interests came into being. Much traffic by water was common. From the few articles still available, quotations follow in hope that someone will remember and enjoy the references.

We find it hard to picture how many tugs, sloops, scows, ferries and other types of boats once were common. For example, "Schooner *Westfield* brought a cargo of lumber for Capt. Morton last Thursday" (June 14, 1898). Captain Morton, it will be remembered, was building a store, part of which was to be used as a millinery shop. "Mrs. Alex McInnis arrived here last Saturday for a visit with her husband" (June 21, 1898). Probably no one remembers them especially.

"Mr. W. Zilcher, of Middletown, Conn., is employed by the E.M.S. Co. as blacksmith at Plant No. 2. Mr.

Zilcher is an expert machinist and will be employed as such when the new plant is in working order" (June 21, 1898). We hope he enjoyed the summer here. "One of the most popular bosses at this place is Mr. Simpson who has charge of a large crew of shovelers. Mr. Simpson has been a boss in the quarries and on large contract jobs and understands the handling of men to the best advantage" (June 21, 1898).

How worried Mr. Jernegan and Mr. Fisher must have been as time went on. It must have been increasingly hard to round up enough gold to keep production to its accustomed level.

According to the account written for the 1976 history of Lubec:

> *Eventually Mr. Fisher ran short of gold dust as a result of largesse in giving away samples to prospective investors. To buy more gold in amounts necessary to keep things moving would have created immediate suspicion. So, when the shortage first threatened to curtail operations, it was necessary to find a means of reducing the drain. Providentially, there came an unusually heavy rain storm which swelled the waters flowing under the Mill Bridge, and this was given as the reason for small amounts of gold dust in the accumulators. Happily the suckers bought this story and the sale of stock continued to boom.*[12]

They must have had some interesting private conferences. They were not seen in each other's company very much. According to a comment made by Mr. Fisher himself to another confederate, "I must not be seen with J.," he wrote. Coming from the same place, too much familiarity might breed suspicion. So detached, decorous behavior was the rule.

Here is an interesting study of the visitors who kept coming and going all through the season. From the issue of the *Herald* for June 28, 1898, we read, "The following persons connected with and interested in the Electrolytic Marine Salts Company arrived here last week: A.B. Ryan, Pres., Boston, W.U. Usher, Treas., Mrs. W.U. Usher, A.P. Sawyer, Director, Miss Abbie L. Sawyer, Miss Augusta Sawyer, Miss Catherine Hopkins, J.E. Moody (Franklin Snow & Co.), F.F. Morrill, Cashier Amesbury National Bank, Albert C. Titcomb, Ex-Mayor, Mrs. Albert C. Titcomb, Mrs. D.L. Barley, and Mrs. W.A. Usher, all from Newburyport; George H. Allen, Director, Manufacturers National Bank, Lynn, Mr. and Mrs. Harry Woodard, Lynn, Chas. Bradley Jr., Providence, C.E. Tullar and wife, Melrose, Rev. Nelson Edwards, Rye, N.Y., A.G. Webster (David Webster Co.), Boston, P.F. Jernegan, Vice Pres. and Inventor, Boston, Robert E. Burke, City Solicitor, Newburyport, W.E. Cusick, Ex-Postmaster, Newburyport, and Louis G. Brockway, Lynn."

Probably these people all came on the Boston Boat. What an interesting time they must have had plying Mr. Jernegan with questions. He must have got so that he almost believed in the project himself.

"The amount of bullion shipped to the U.S. Mint, New York, in 16 shipments, was a total of 2,877 ounces, valued at $20,000. The company has declared a dividend of three percent on the stock issued…500,000 shares…There are working, at present, 239 machines, producing $239 per day net profit. A plant is now in preparation that will contain 5,000 machines. It is expected that part of the machines will be in operation by November."

On July 5, "Mr. A.B. Ryan, Pres. of the E.M.S. Co., returned to Boston." It kept him busy traveling. "While all this was going on, Mr. Moody Knight is working his double team here for Mr. Shanahan," it was noted.

"Mr. Frank Brown, working for Mr. Otis Brown, cut his leg, badly with an adze last Friday" (June 5, 1898).

At the Canal, July 5, it was announced that "an extension is being built on our wharf to accommodate

A circa 1900 photo of the Eastport–Campobello–Lubec ferry wharf in North Lubec as it appeared after its extension. The ferry conveyed passengers from the Eastern Steamship Company dock in Lubec Village, where potential investors from Boston and New York arrived. *Courtesy of Pauline Bailey.*

the boats of the Passamaquoddy Ferry Co. These boats, it is understood, will make regular trips here each day after July 4." Who recalls if they did so?

BUSINESS ANGLES OF THE E.M.S. CO.

Article 39, September 8, 1949

The Electrolytic Marine Salts Company was organized in Portland, Maine, in the Law Office of Levi Turner, but the main office was in Boston, at 53 State Street, in the Exchange Building, Room 502. It was interesting to call in at that historic room, now occupied by a firm of lawyers. Here, much of the drama of the company's dealings occurred. [Built in 1891 as the Boston Stock Exchange, the façade remains attached to the 1984 forty-story Exchange Place.]

Following is the beginning of an article taken from the *Boston Weekly Journal* of Friday, August 19, 1898. This was, of course, about two weeks after the disclosure of the nature of the gold "Experiment."

"Marine Salts Directors Hold Meeting.

"There was a crowd of discontented people in Lorimer Hall, Tremont Temple, Tuesday morning. It was a meeting of stockholders of The Electrolytic Marine Salts Company of Boston. It was not a legal meeting, as the corporation is a Maine establishment, and the bylaws require that all meetings shall be held in Portland Friday afternoon at two o'clock, and to outline the views of the stockholders here who may not be able to attend the Portland meeting.

"President Ryan called the meeting to order at 10:15, and asked for the nomination of a chairman. The hall was two thirds full, but for some minutes, no one spoke. Then attorney Kelly of Haverill nominated Robert E. Burke of Newburyport. He declined, as he had a report to make.

"Then Hon. Joseph O. Burdett was named and elected. Mr. Burdett was counsel for several stockholders. He announced that he represented a stockholder holding a thousand shares. This was a piece of information defining his position in the meeting. Nathaniel N. Jones of Newburyport was nominated for secretary, but declined, and Charles T. Patten of Worcester was elected to that position. Mr. Burdett said it was desired to know what position the stockholders wanted to take regarding the action of the Directors at Portland, Friday, and as to winding up the corporation. He related the action taken at the last meeting, and the appointment of Mr. Burke to go to Lubec and investigate the plant there. He asked that Mr. Burke report.

"Mr. Burke then reported. He detailed his instructions as to investigations at Lubec. He said he went there a week ago Friday. He got back Sunday. He had not attempted an investigation at the Boston end. Since he had been at Lubec, he had been engaged in preventing litigation involving the company suits and dealing with claimants wanting their money. From $3,000 to $5,000 would settle all claim of the manufacturers of the accumulators."

MARINE SALTS DIRECTOR MEETING

Articles 40 and 41, September 15 and 22, 1949

"There was an order for 4,000 new machines at $50 each. The manufacturers agreed to compromise for $16,000. The manufacturer was Mr. Ball and this sum he claimed as damages by breaking of the contract. Electrolytic Marine Salts Company. The detailed statement of Mr. Burke was as follows:

"Report of Attorney Robert E. Burke.

Received from Treasurer to Aug. 10 from beginning of business $88,025

Accounts sent to Boston office for payment $40,972

Ledger accounts we owe (understood $3,000 been paid on this at Boston office) $6,328

Received cash discounts on bills paid $47

Amount due plant $649

Total expenditures $136,023

Expenditures:

Bought lumber and piling $48,073

Paid for construction $33,888

Labor E.M.S. Co. $19,271

Materials $10,023

General-expense account, including engineering department, Insurance etc. $4,052

Bridge at Lubec $3,784

Boats $1,370

Fuel purchased $1,119

Tools purchased $1,048
Real estate etc. $586
Stable account $258
Furniture and fixtures $287
Machinery, hoist engines etc. $210
Reservoir account $175
Vose & Reynolds, paid on account $2,000
Cash in Bank $12,070
Cash in office $2,102
Total $136,023

"Inventory of Property at Lubec
Lumber, piling, lathes etc. $9,442
From Town of Lubec on Bridge $1,500
Boats $1,331
Fuel $600
Materials $2,961
Furniture $158
Horse, wagon etc. $130
Buildings $450
Real estate $5,000
Plant No.1 $10,630
Total $46,538

"It was moved and voted that there be appointed an Advisory Committee of five to act with the Directors in the control of the Company, and, if in their judgment best, to make such test of the process as they saw fit. The motion was amended to provide for the appointment of this committee from the floor. An effort to divide the question was unsuccessful. It was voted, however, that the committee report within 30 days.

"The motion for the appointment of a committee of five to act as an advisory body to the officers in control of the company, and to make any test of the secret process thought advisable, was then carried.

"Mr. Kelly of Haverill then offered a motion that the committee be instructed to make no further tests, either at Lubec, or anywhere else. The Chairman ruled the motion out of order, as the meeting had but a moment before voted otherwise. Then Mr. Kelly moved a reconsideration, but meanwhile, a weary stockholder had moved an adjournment, and this was seconded and carried with a large spirit of unanimity.

"It was stated during the meeting that there were 10,000,000 shares of stock of which Jernegan and Fisher personally owned or controlled 7,000,000."

This concludes the account of the Directors Meeting in Lorimer Hall, Tremont Temple, as printed in the *Boston Weekly Journal* of Friday, August 19, 1898.

It will be noticed that the sum of the figures is not, by any means, the correct total. It is, however, copied exactly as given in the paper. We will have to speculate as to who made the error.

More Angles of E.M.S. Co.

Article 45, October 20, 1949

At the risk of repeating some of the information already published about the E.M.S. Co., the following article, taken from a publication called the *Beacon*, sets forth a few angles that are different. This article was printed rather early in 1898, it would seem.

"Gold in Quoddy What the Yankee Theologians Intend Doing at Lubec [The word "Quoddy" is short for Passamaquoddy, the anglicized name of the indigenous Native Americans still resident in the area. Passamaquoddy Bay, part of the Bay of Fundy, lies along the eastern shore of Maine west of Campobello Island, New Brunswick, out to "Quoddy Head," the easternmost point of the United States. "Quoddy" thus refers to this region.]

"A week or two ago the *Beacon* called attention to some mysterious works that were being carried on at

North Lubec by a number of Western capitalists. It now turns out that the scheme, which was so secretly guarded, was the gathering of gold from the waters of Quoddy by the means of electricity.

"The two originators of the enterprise are A.B. Ryan of Middletown, Conn., superintendent of a Baptist Sunday School at that town, and Rev. Prescott F. Jernegan, pastor of a Baptist Church in Deland, Fla. The chief agent of the company at Boston is the Rev. C.E. Tullar, of Melrose, who is pastor of one of the prominent churches at that town.

"It is too long a story to tell how these theologians became interested in the subject–but interested they are and very deeply too. The exact electrical process by which the precious metal is to be drawn from the water is kept religiously secret, but all their plans have not been so well guarded.

"President Ryan, in talking with a newspaper man on the subject, said, 'We plan to build such apparatus as will ensure us an income of $100 to $150 a day. That means that we shall be able to handle about 4,000 tons of water., It will be easy. Passamaquoddy Bay is filled with inlets suitable to our purpose. With the proper rise and fall of tide we shall have no difficulty. For instance, we build our apparatus across the mouth of one of the coves. The tide in the Bay falls about twenty feet. That will give a speed of about six miles an hour, and for falling six hours, that gives us thirty six miles of water every tide to work on. In running out, the water passes through our apparatus. Through chemical action it is made to deposit gold–not all the gold, but some of it: enough, however, as I say to ensure us an income of more than $100 a day. Even if the same water came back on the next tide it would still serve our purpose. It's good for more. We already have two small mills at work in one of the inlets, but we are saying nothing until the $50,000 Plant is working. Then you will open your eyes. The Plant will be in full operation the first of the year, now that sufficient capital has been promised us.

"Our reasons for locating the Plant at Passamaquoddy Bay are obvious. First, the Bay provides secrecy. Of course the success of the scheme depends on our secret. Should that get abroad we would be lost. That is why we have not patented the invention. We intend to have the sole rights. If we should get it patented, someone else could go to the patent office and copy it. It would be easy. Then he could go to some other part of the world and start to work. It may be necessary for us to do something like that ourselves. The very fact that our invention is not patented, and the secrecy of the work and jealousy, may provoke an investigation by the government. Then we might be obliged to give up our Plant on Passamaquoddy Bay. But we would have the rest of the world for a field. Oh, we are well-protected.' Mr. Ryan says a Boston Capitalist offered to buy the discovery outright for $100,000 but the offer was refused."

There are some funny thoughts in this description in view of the fact that Mr. Ryan only thought that he knew the "secret" of the working of the plant. Indeed, it was necessary for some of the organization to go to some other part of the world to work. And wouldn't it have been intriguing to have had a government investigation in Mill Creek fifty years ago?

CHAPTER 6

A Klondike Cast of Characters

About Mr. Prescott Jernegan

Article 4, January 6, 1949

Although the promoters of the Electrolytic Marine Salts Company appeared to be able and honest, to thinking people it did seem peculiar that two amateurs should "stumble on the secret" and accomplish what scientists had failed to do, namely extract gold from water. Mr. Jernegan was, however, reputed to have an exceptional interest in chemistry, but one recounter was mean enough to suggest that it was alchemy instead.

Mr. Jernegan himself said, "Fisher knows a lot about electricity without being exactly an electrical engineer." So, partly with Jernegan's chemistry and Fisher's electricity, the "secret process" came into being.

In fact, Jernegan is recorded as stating that he received his ideas on this fascinating subject "in a vision." When a brother minister, Reverend Charles A. Piddock, heard this, he said, "So far as I was concerned, that settled the matter." Pastor Piddock accompanied this comment, according to the *Hartford*

Prescott F. Jernegan, circa 1892, believed to be about the time of his marriage. *Courtesy of the Lubec Historical Society.*

Courant, with a "glorious guffaw." (Reverend Mr. Piddock had been a predecessor of Jernegan's in the First Baptist Church in Middletown, Connecticut, and was later in Hartford.)

To delve into Prescott Ford Jernegan's life a bit more, he belonged to a respected family of Edgartown, Martha's Vineyard. His father, Captain Jared Jernegan [1825-1899], had made a fortune in the whaling business but suffered reverses when his fleet was wrecked in the Arctic about 1883. There was a brother, Marcus, and other children in this family of Captain Jared's.

Prescott was exposed to a good education at Phillips Andover Academy [Andover, Massachusetts], Brown University [Providence, Rhode Island] and Newton Theological Seminary [Newton Centre, Massachusetts]. At one time he taught at Andover. There was also a period of European travel as companion to a wealthy invalid.

As his wife, he chose Miss B. Eveline Phinney, also of Edgartown. Miss Phinney was the daughter of Captain Adelbert Phinney [Sandwich, Massachusetts, 1844–Seattle, Washington, 1928]. They had one son, who was a small boy at the time the Jernegans lived at North Lubec. It might be wondered what turn his fortunes took.

As a minister, he was considered somewhat erratic by many. At least he was different. He served at Middletown, Connecticut, from March 1892 to July 1895. According to a church history written in 1945, it is stated, "More than any other pastor, he preached a social gospel. In spite of meager resources he opened a wood yard on lower Court Street, for the employment and rehabilitation of homeless men. Men of the streets were given temporary shelter at the Parsonage in the hope of assisting them to face realities more constructively. When one beneficiary stole silverware given him as a wedding present, he declined to prosecute."

In addition to inaugurating the Industrial Woodpile for tramps, Reverend Jernegan is remembered for other ideas. One Sunday he announced that he would henceforth serve "without money and without price," but if his work among them had merit, depend upon their generosity. It developed that he did not think highly of the generosity of the congregation, and, as often, his methods did not please.

During his term as pastor, the "sittings," or pews, were made free.

The following title to one of his sermons shows originality: "The Blessedness of Being Misrepresented and Lied About."

After dissension concerning the question of salary, Mr. Jernegan accepted the pastorate of the First Baptist Church in DeLand, Florida, and Mrs. Jernegan returned to Edgartown.

Much was made of the fact that Mr. Jernegan was a minister, but as Reverend Piddock observed, "C.E.

Fisher, who seems to have been at least the equal of Jernegan in the conspiracy, is High Church Episcopalian, a member of the church in Brooklyn." He then added that there are scoundrels in all denominations. But that was fifty years and more ago.

ABOUT CHARLES E. FISHER

Article 5, January 13, 1949

Charles E. Fisher, like P.F. Jernegan from Martha's Vineyard, had led a varied life, and it was said that he finally had a desire to settle down. This was at the time of his last-known venture at North Lubec, where his aims were said to have been 1) to get his debts paid and 2) to invest enough money so that he might live on $^{2}/_{3}$ of the income.

Mr. Fisher had been a one-time student of medicine. He had served in the British army. He had worked in a New Bedford factory and had been a floorwalker in a department store. In addition to all this, he was an excellent professional diver, which accomplishment he did not publicize during the inception and life of the Electrolytic Marine Salts Company.

By all odds, Charles E. Fisher "put on the best show" as an officer in this company and in his manner of daily living. Only recall how impressive, not to say magnificent, some of those floorwalkers can be. Coupled with an ability to judge people, it is small wonder he conducted himself with assurance. He dressed with great style, a dude they might have called him in 1898. He was of average size, lithe and exceedingly personable. Pictures of him show thick, lively hair and a flourishing handlebar mustache.

Mr. Fisher owned saddle horses, which were objects of marvel to the small boys of the neighborhood.

Lawrence Brothers Store and the Lawrence Cannery, built on pilings over water, circa 1906. *Courtesy of the Lubec Historical Society.*

Lawrence Brothers Store, December 4, 2012. Compare with the circa 1906 photograph. The two houses to the north (left) of the store remain, largely unchanged, as do the side door and small window on the store's basement and the fire hydrant across North Lubec Road. *Photograph by Ronald Pesha.*

These horses were kept in a barn that was then located across from Lawrence Brothers store. In company with various ladies of his household, he would go riding about town.

There was a good "cellar" at the Fisher house, it seems. Once when some local workmen were busy there, they made bold to help themselves and were pleasantly surprised not to be reprimanded.

At Plant No. 1 (the old gristmill that had been remodeled and enlarged) was the laboratory to which no one, not even the other officers of the company, was admitted without a pass from Fisher. Once someone asked Fisher if there was not danger that the employees might not find out the secret process. He said that a person could do laboratory work and still not know of the process of extracting the gold. How right he was.

There was speculation among scientific men how Fisher, not an engineer, could have deceived so many people with his bogus devices. It may be repeated that he was enhanced with a personality that fairly exuded confidence and that he had remarkable persuasive powers.

To show what diverse things were printed at the time, one article, highly colored and erroneous, characterized Fisher as "slight, nervous and shifty eyed," and Jernegan as "big, ungainly and boyish." At all events, a collaboration resulted between these two forces and according to many who ought to know, Fisher had the dubious distinction of "manipulating the whole business and fooling the best of them."

Others at Plant No. 1

Article 59, January 26, 1950

It appears that Fisher, himself, was not introduced to the company at the very beginning. Jernegan told Mr. Ryan that he must have in his employ only men that he could trust implicitly; men whom he knew. He then spoke of his friend Fisher and insisted that he was engaged because of their joint interest in the discovery. Thus he came into the scheme.

Mr. Ryan also had this to say: "Fisher mystified the officers with stories of himself. He told me that he was once in the English and American armies, that he was once a submarine diver, and also that he had been a detective, once upon a time. He also said

Sardine workers' houses or factory camps on the North Lubec Road, circa 1950s, owned by the Lawrence Brothers North Lubec Manufacturing & Canning Co. The view is north up the road, in the area of the Roger's Island parking lot, with the home of Carrie Comstock Bangs, who wrote most of the text in this book, visible in the far background. Andrew Pierson, "Klondike" superintendant, owned and occupied the residence during the Jernegan era. *Courtesy of Tom Dean.*

that Phelan had been a celebrated counterfeiter, and was known the world over for his marvelous success in that line.

"Jernegan, Phelan and Fisher were noticeably upon excellent terms and frequently in each others' company while in New York. When the letters which passed between Jernegan, Phelan and Fisher are taken into consideration, the officers of the company think they can understand who are referred to by the heretofore mysterious initials in these letters, which are claimed to be evidence of the crookedness of the scheme."

Well might the officers have given pause with all these peculiar actions.

A few articles back, mention was made of the death of Fisher in New South Wales in 1900. It is commonly believed that Fisher, himself, put this notice of his own death in the papers. Then he was free to roam about without so much danger of detection. Far from being interred in Australia, where could he have gone.

About Andrew N. Pierson

Article 7, January 27, 1949

Andrew N. Pierson was a florist from Cromwell, Connecticut. Cromwell adjoins Middletown, where the Reverend Jernegan had held a pastorate. Mr. Pierson was superintendent of the activities connected with the operation of the project of the Electrolytic Marine Salts Company.

He was of Scandinavian ancestry, a large and very attractive man with an enormous grasp for detail and the vigor for putting ideas into action and transplanting plans into realities. Being convinced from observations of the results of experiments that

The Bangs House. Before it burned, the house was on the east side of North Lubec Road, today's address about number 579. This drawing appears in the 1976 book *200 Years of Lubec History, 1776–1976. Courtesy of the Lubec Historical Society.*

the process of extracting the gold from seawater was practical, Mr. Pierson was among those who furnished capital for the enterprise. So complete was his faith in this work into which he threw himself wholeheartedly that he himself bought the company's stock extensively and interested many of his Swedish friends in Cromwell in doing likewise. Moreover, about thirty men from Cromwell came to Lubec to obtain employment on the project.

Unaware of the duplicity that was being practiced by the Messrs. Jernegan and Fisher, A.N. Pierson's belief in what he was doing and his obvious respectability were such as to inspire confidence. He was a man whom everyone liked, and stories of his kindness, greatness of heart and practical Christianity are legion.

A.N. Pierson bought the house, then new and not entirely finished inside, from Albert Leon Bangs [1865-1946] and completed it. Two underground drains were constructed across the yard in front of the house. The driveway was filled in with a subsurface underpinning of stone so that no grass could grow there. He installed a gravity water supply into the house, piping the water from a couple of deep wells on the hill across the road. The plumbing was put in by E.E. Trecartin, and this private system gave satisfactory service for more than forty years.

In line with his interest in the florist business, Mr. Pierson found time to have a small green house built on the rear of the house. Mr. J.D. Huckins was among the local carpenters who worked on this building. This "glasshouse" resembled a sunporch of today. It was surrounded by sliding windows, and there were oblong and triangular zinc-lined trays around the inside walls. The sides of the roof [were] curved, Chinese fashion, and were later productive of copious leaks. The top of the roof surrounded by a railing was flat and covered with "tin." Such a tattoo as a heavy rain would pound out on that roof. In this hothouse, Mr. Pierson held a Sunday school and (we hope we are right) occasional evening song services. Mr. Moses Pike Lawrence recalls attending the Sunday school. It would seem that Mr. Pierson was doing the work that might have been expected of Mr. Jernegan. A.N. Pierson really believed in his mission and entered actively as a citizen into the life of the community where he sojourned. He might be likened to those "who, passing through the valley of Baca, make it a well."

Mr. and Mrs. Pierson had three children, Frank, Wallace and Emily. Many interesting people visited at their home while they were at North Lubec.
On one occasion, while Mr. Pierson was about his horticultural pursuits, someone asked him why he did not paint the flowerpots gold. He laughed and made some pleasantry about "covering everything with gold." Simply hundreds of those small clay pots were left after Mr. Pierson's departure. Besides being

used for practical purposes, children liked to play house with them. In a recent year, the last of these flowerpots were presented to Mr. Arthur Tyler.

To go about his work, in accordance with the custom of the time, Mr. Pierson had a horse and wagon. In front of the house, for many years, remained the wooden hitching post with an opening at the top like the eye of a needle.

When the curtain rang down on the Klondike affair, Mr. and Mrs. Bangs returned from Buck's Harbor, where they had spent the intervening month, and bought back the house complete with improvements for the same amount for which it had been sold.

A few years before Carrie Bangs wrote this extensive newspaper series about the "Klondike," Bill Aye recalled the Klondike gold extraction scheme in the *Lubec Herald* on April 4, 1946:

> *Pierson became the manager of the company, Jernegan being the promoter. There was another accomplice by the name of Fisher, who was in the know, and these two were the crooks. All others connected were honest men but duped like the stockholders.*
>
> *This Mr. Pierson became manager of the company. He arrived here on the Boston boat October 6, 1896, with a chest of tools, a small donkey engine and some other articles which were to be the beginning of the Electrolytic Marine Salts Co. On inquiry at the boat office as to where he could get transportation for himself and his implements, he was directed to F.S. Reynolds by the agent as the man who could furnish all the teams he wanted, so forthwith F.S. was engaged on a project which kept him on the hustle for the next year and a half.*
>
> *There must have been some advance preparation which no one around here knew about for this man Pierson went up to W.J. Mahlman's and bought it right out, and the next day had men out cutting logs to make a wharf so they could operate their gold separation boxes which had to be breaking water at low tide.*
>
> *Everyone that came along was hired at good pay, and as the amount of logs that they were getting out of North Lubec were not near enough to supply their wants, F.S. Reynolds was contracted with to supply a larger amount. He had*

This rare old photograph from 1898 shows actual construction of a wooden dam across the Canal's entrance to impound water in Fowler Basin to the west, where another dam was located. *Courtesy of the Lubec Historical Society.*

> *schooner after schooner load brought here from New Brunswick and the upper part of the bay was like a drive on the Kennebec River, logs and more logs. They also engaged a contractor by the name of Shanahan who was putting up the railroad dike at Machias. He had a large crew, mostly Italian laborers, who later on figured in the picture when the scheme folded up.*

Now we return to Ms. Bangs's narrative about Andrew Pierson.

Through the kindness of Mr. and Mrs. Fred Trecartin, some parts of the *Lubec Herald* for 1898 have been available. The following items appeared in the June 1898 issues under a weekly column that was entitled "Nuggets from the Canal."

"Mr. A.N. Pierson, manager of construction at the E.M.S. Co.'s Plant No. 2, is one of the busiest of men. He seems to be everywhere at the same time. His constant supervision and careful investigation of all work under way results in steady and rapid progress all along the line. Mr. Pierson's master hand guides all the work at the Canal and his example and spirit seems to dominate the action as well as the spirit of his workmen.

"Mr. Frank Pierson, oldest son of A.N. Pierson, Esq., is visiting his parents at North Lubec. Mr. Pierson the younger is an enthusiastic sportsman. We can promise him some sport, deer hunting if he visits Lubec next fall." ["Pierson was the 'good guy' of the gold hoax promoters, while the bad guys were the erstwhile minister and school teacher, Prescott Jernegan, and Charles Fisher, whose job it was to plant the gold," wrote Olaf A. Bangs (1908–1987) in the 1976 book *200 Years of Lubec History, 1776–1976*. Son of Albert L. and Mary C. Bangs, Olaf was the younger brother of Carrie C. Bangs.]

"Mr. A.N. Pierson, manager of construction at Plant No. 2, is having the Good Templar's Hall (Canal Hall) repaired and put in condition to be used as a meeting house. Mr. Pierson has been holding cottage meetings at his home here, but finds the interest growing and that his house will not accommodate the audience, consequently he has taken the hall in hand and will make an excellent meeting house of it. Right here, we wish to state that as the time passes, Lubec will find in Mr. Pierson it has gained a citizen who at all times has the best interests of the community at heart. In his own town he had always been noted for his liberality in all good works, and has done more than anyone else to build up his town. Many stories of his kindness of heart are told by those who know of his generosity."

Jernegan's Father, Captain Jared Jernegan

Articles 43 and 44, October 6 and 13, 1949

A sketch of Captain Jared Jernegan, the father of Prescott Ford Jernegan, shows him attired in working clothes and a cap such as used to be worn by seamen. For Captain Jared was one of the noted Arctic whaling captains. The picture shows him to be a sturdy and powerful man.

In an account of Jernegan's ancestry in the *Boston Sunday Globe* of August 7, 1898, we find that "the facile inventor" of the process of taking gold out of the public pocket, passing it through crucibles along with mud from salt water, and into his own purse, comes from a long line of honest men. His forefathers have for generations been numbered among the solid citizens of the quiet little seaport of Edgartown, on the southeast side of Martha's Vineyard.

"They followed the sea, and had respect for salt water. They were not given to great show of religion, being too busy in toiling for the means wherewith to maintain an honest estate. The first Jernegan to come to Martha's Vineyard, William, who won the prefix 'Honorable,' for his name, was an immigrant from England to Virginia, and finally came up the coast and erected his roof tree on the pleasing old island off the Massachusetts southern coast. That was when the last century was still young. He fought in the Indian Wars, and was heard in the councils of the colony.

"Since then, the Jernegan line has been unbroken in Edgartown. Its honored representative there at the present time (1898) is Captain Jared Jernegan, father of Prescott F., a hale and hearty old mariner who is passing his waning days in his comfortable home on a shady side street of the drowsy old village where he was born, and from which he sailed forth for more than fifty years to the south seas (and north) for whales.

"When Prescott was still a child, he was taken on a voyage on his father's whaling bark, the *Roman*, along with his mother and older sister. On the voyage, there was a mutiny aboard the ship, and only the coolness of Captain Jernegan averted bloodshed. Had the mutinous whalers succeeded in their designs on the life of the captain and his family, it is easy to see that there would not have been spared to posterity the interesting person of Prescott Jernegan.

"The boy was about two years old, when this episode in his life occurred. His father's vessel was at anchor in a roadstead [a nautical term for an anchorage less sheltered than a harbor] in the Marquesas group of islands in the South Pacific, when seventeen of the crew of thirty-eight men rose and refused duty in getting the ship underway. They knocked the mate down and shut the captain up in the cabin.

"'They said they were going to kill me,' said Captain Jernegan, relating the story. 'I told them to try it. I had my rifle aboard, and I got it. Then I told my wife and the children to get into their stateroom. I stood guard with the rifle. When the men found they could not frighten me they got into boats and left the ship for the island. They had got hold of some liquor and went ashore for some more. While they were gone, I slipped the anchor and stood out to sea, leaving them there.'

"Captain Jernegan never took the boy on another voyage. The lad passed the rest of his tender years in the fostering care of his mother at home."

The following account from the *Boston Globe* of August 7, 1898, shows how Captain Jared Jernegan, the father of Prescott Ford Jernegan, received the news of the failure of the gold-from-seawater venture. Shortly after the exposure, a reporter called on the elder Mr. Jernegan, and here is what he had to say:

"The writer found the old man in his market garden, up the street a little from his home, where he delves in the ground to keep his naturally industrious bent satisfied. While bearded, but still with a clear skin, and twinkling eye, Captain Jernegan came out of the garden gate, carrying a pan of silvery green corn, ready husked for dinner, when he learned the visitor's mission.

"He put down the pan of corn and replied to an expression of interest in the trying position in which his son had placed him. 'He never mentioned gold to me,' said the old captain. 'He knew better.' I had sailed over too many miles of salt water to be taken in. He never confided his plans to me. If he had, I would have told him to stick to the truth at all times. My father told me when I was a boy to always tell the truth if you don't save a cent. I stuck to that advice. My boy, Prescott, has gone wrong. He did not follow the teachings I tried to give him. Now he must take the consequences. I shall not worry about him.'"

"The old man looked his visitor straight in the face. His bright eyes told a story of fatherly love not entirely hidden by his Spartan fortitude. 'They said here, the news would give me a shock,' he continued, 'Shock! It would take more than that to give me a shock. I am able to hold up my head and say I always treated every man square. If my son can't say that, it is not my fault.

"'When my boy Marcus came down here with that valise full of money his brother Prescott had given to him, I said to him, don't keep a cent of it. Give it all up. Tell the truth and I will love you. He turned the money over to the detectives. It was for some of the people who invested. I think a minister and somebody else. Anyway, we had no right to it, and Marcus gave it up. Marcus is all right. He was honest in helping his brother. He has done all he could to make things right, and I honor him for it. If Prescott had followed my advice he would not be hiding now.'

"Capt. Jernegan picked up the pan of corn. The sunlight filtered through the branches, fell athwart his kindly old face. 'The women folks take it pretty hard,' he said. 'His mother is all worked up over it. If they wasn't I would ask you in.'

"'You can say for me,' he went on, 'if you say anything, that I am ready to face the world in spite of it all. My record is clear, and so was the record of my forefathers. No man can say I ever failed to meet an obligation. You can say that for me.'

"And drawing himself up, the old man again sought to hide his sorrow behind a brave front. There was something touching in his fine old figure as he stood in the street, holding the pan of corn from his little garden, that compelled a thought of his fugitive son, at that moment fleeing across France loaded down with the proceeds of his wonderful process of extracting gold from sea water."

Well, that is a beautiful account and it should show us that it is not only ourselves that we hurt when we embark on questionable ventures.

Jared Jernegan died about six months later, on January 13, 1899, at age seventy-three. Prescott's mother, Helen, died on February 26, 1934, at age ninety-six.

More Sidelights on Fisher and Jernegan

Article 57, January 12, 1950

From the *Boston Herald* of Wednesday, August 3, 1898, while the story of the seawater fraud was at its height, a reporter gave a summary of the early lives of Fisher and Jernegan. We have seen that they were born and brought up in Edgartown, Martha' s Vineyard. The article states that "they were boys and playmates together. They had often schemed, planned, and executed in mutual ventures in their youth.

"Fisher's father was an honest man. He was the head of a company which extracted whales from salt water, without the aid of a submarine diver or a 'secret process.' His works were not called electrolytic, but a common everyday whaling vessel, and the senior Fisher was the captain. He employed neither electricity nor chemicals. He used the harpoon and the fire to extract his gold from the great whales which he dragged up from the salt water.

"Young Fisher did not enjoy the frank, open process of his father, so, early in his youth, it is said, he set out to get his gold by an easier and a secret process. He went out into the world and saw how others got their gold. He went to New York and other large cities." [A few days later, the *Boston Globe* expanded on the *Boston Herald*'s reportage.]

THE BOYHOOD OF CHARLES FISHER

Article 46, October 27, 1949

From a revealing article in the *Boston Sunday Globe* of August 7, 1898, we read of the younger days of those boys of Edgartown, Prescott Jernegan and Charles Fisher. While there seems to have been but little to admire in Fisher, yet we may marvel at his audacity.

Of Jernegan, as a youth, it was stated that "he grew to be a long ungainly boy, of rather hulking appearance, and of serious demeanor at most times."

"Fisher was his opposite in both appearance and temperament. He was quick, nervous, and in his moral status what Yankees call 'shifty.' He did not care specially for books, and never took any prizes at school for either deportment-ship or scholarship.

"In the winter of 1881, when Fisher and Jernegan were about 16, there was a religious revival in Edgartown, and to this event may be traced the development of Jernegan's religious profession. The revival was one of the sort that stirs a whole community. It was conducted by the Reverend L.T. Johnson, a highly successful evangelist. The meetings were held in the Methodist Church, and so interesting did they become, that the services at the other churches were entirely eclipsed, and those of other faiths flocked to the old pillared edifice on Main Street where the fervid preacher was claiming converts nightly. 'The whole town was 'getting converted,' and Jernegan and Fisher naturally went to the meetings.

"In a short time, Jernegan rose for prayers. He became an earnest convert, and it was not long before he was expounding the faith of saving grace to his companions. Fisher also rose for prayers and expressed his unworthiness and hope of salvation. Jernegan soon joined the Baptist Church which his family attended. Fisher's conversion was not permanent and he soon fell back into the ranks of the unregenerate youth of the town.

"As the boys grew up, their ways parted. Jernegan, after leaving the Edgartown High School, went to Phillips Andover to prepare for college. Capt. Jernegan was then reported to be well to do, and he could afford to give his boy a good education. He has since lost much of his modest fortune, through Arctic disasters to the whaling fleet.

"Fisher remained in Edgartown, and after leaving school, went to work. He gravitated toward the mainland, and got a place in a mill office in New Bedford. There he made a good impression. From that time on he went from one occupation to another. At one time, he was a floor walker in a Brooklyn dry goods store. At another, he was working as a submarine diver. This was the trade that made him invaluable in the gold making business, as without it, the gold "machine" could never have been loaded under the waters of the Narragansett Bay while the unsuspecting Messrs. Ryan and Pierson of Middletown, Conn. were waiting for the precious metal to accumulate in the sunken box.

An Edgartown publication by the then Dukes County Historical Society characterized Fisher as a soldier of fortune, "dashing, clever, bold" with a "gentle disarming manner" characteristic in popular culture of the confidence man implanting credence among the gullible.[13]

"Nobody in Edgartown pretends to know the many goings and comings of Fisher. He was often at home, and he had the reputation in the town of being rather a gay boy. There was one period when he did not return to his native isle for three years. In that time he was in the English Army as a trooper. Finally, he came back a handsome dashing fellow, suave and interesting, and altogether, a smooth "rolling stone."

What experiences may he not have had after leaving Lubec in his early thirties.

The Early Life of Jernegan

Articles 47 and 48, November 3 and 10, 1949

The Boston Globe of August 7, 1898, in reviewing Jernegan's career up to that time, sheds light on the things that made him what he was. Much that was deplorable must have been inborn, and certainly not the fruits of his upbringing. To quote, "This is the life story of Jernegan, Rev. Prescott Ford Jernegan, erstwhile preacher of the gospel, and later spinner of golden meshes from the water wherewith he entangled the worldly, whom, as a fisher of men in the paths of righteousness, he had sought and found in their ways of guile.

"His net, thought not in fact as weak as a rope of sand, caught so many fishes, big and little, that the originator deserves a place all by himself among those who have wrought for similar ends.

"While something of a Col. Sellers, Jernegan far outshone his prototype of the drama. [Colonel Mulberry Sellers was a character in an 1873 novel *The Gilded Age: A Tale of Today*, written partially by Mark Twain, and subsequently in an 1874 play titled *Colonel Sellers*. The comic charlatan engaged in assorted swindles, seldom successful. The novel is credited with introducing the phrase "Gilded Age," derived from a line in Shakespeare's 1595 play *King John*.] Col. Sellers built his castles on eye water for the African natives, and dreamed there were 'millions in it.' Jernegan built his on sea water, and took millions out of it, or, if not millions, at least enough thousands to make travel very worth while."

The life story of Jernegan is not one of thrilling interest or varied incident. Up to the time of his great coup of wringing gold out of the sea before the eyes of admiring stockholders in his company, he never did anything extraordinary, though he was a brilliant student and speaker. If a study of his life proves that he stumbled on the peculiar possibilities within him, without realizing, at first, what a creative genius he was.

"In his youth, Jernegan was not a particularly marked boy. As a young man, he threw himself with zest into what he considered religious thought, now shown, however, to have been not unmixed in later life with thoughts of another kind. As a student, he toiled hard at his books. As a preacher, he was exceedingly active, but attracted more attention by ill-considered remarks and acts rather than by his preaching, though that was excellent in its way. As a romancer and manipulator of the greatest hoax of his time, he was destined to come before the world."

In conversation with a native of Jernegan's hometown of Edgartown last weekend, it was surprising to learn how few of the details of the Klondike affair and of the persons most concerned were known. The name of Charles E. Fisher meant nothing to this person. Perhaps all of the family moved away from the island years ago. Mrs. Jernegan lived many years after the death of her husband, Captain Jared, but was seldom seen in public.

In continuing the story of Prescott Jernegan's educational career, we read that "the cloth of the clergy was donned by Jernegan after his graduation at Brown (1889), he being ordained the same year. He first turned to teaching, however, instead of the pulpit, becoming an instructor at Phillips Andover in Greek and Latin."

After that, his friends in Edgartown heard that he was traveling abroad with a wealthy Englishman as a companion. The Englishman was a liberal patron, and as a result, Prescott Jernegan was in funds. Finally he came back home, and the town learned that the Englishman, the good friend of their serious-minded young townsman, was dead. Edgartown is now asking itself if the Englishman was not as much myth as Sarah Gamp's Mrs. Harris. [Sarah Gamp, a fictional character in *Martin Chuzzlewit*, written by Charles Dickens in 1843–44, was given to exaggeration and fabrication.]

"Jernegan's travel did not divert his attention from his declared purpose of becoming a divinity student, and in due time, he went to Newton and began to study in the theological school there [Andover Newton Theological School, located in Newton Centre, Massachusetts]. Here, as at Brown, he had the reputation of being a tireless student. He was graduated from the Theological School in 1892 and was highly recommended by the faculty when a committee from the Baptist Church in Middletown, Connecticut, came there in search of a likely young preacher to fill their pulpit.

"That committee was headed by Mr. Arthur B. Ryan, a Middletown jeweler, formerly of Hartford, and now (1898) president of the defunct Electrolytic Marine Salts Company. The stars that governed the fate of Rev. Prescott Ford Jernegan were approaching conjunction the day Mr. Ryan and his fellow church workers journeyed to Newton from the valley of the Connecticut. From then on, events leading to the end were more or less rabid. Mr. Ryan discovered Mr. Jernegan. It did not take Jernegan long to make use of Ryan.

"Jernegan went to Middletown in 1893. He was then a finished preacher as he had supplied pulpits in various places during his vacations and after his graduation from Newton. He preached two or three sermons at the Middletown Baptist Church to show his quality, and as a result, he was engaged as pastor with a salary the first year of $1,400."

CHAPTER 7
Minister in Middletown

The Middletown Pastorate

Article 49, November 11, 1949

There are but few cities in New England more beautiful than Middletown. Its location in the great valley of the Connecticut is charming. It is a city of wide, clean streets, shaded with trees that cast their shadows over handsome homes. The town is crowned with the buildings of Wesleyan University. [Some sixteen miles south of Hartford and considered part of that Metropolitan area, Middletown lies on the Connecticut River and, as of 2013, boasted a population of about forty-eight thousand.]

"The main business street of Middletown is parallel with the river. It is wide and straight. Most of the churches in the town are on it. The Baptist Church, to which Reverend Prescott Ford Jernegan was called, is not one of the richest in town, but it has a congregation of estimable people who have ideas of their own as to how they want the gospel preached to them. The church was originally an offshoot of the South Congregational Church of

First Baptist Church, Middletown, Connecticut, in 2012, Reverend Jernegan's pastorate from March 1892 to July 1895. *Photograph by David Bauer, Middlesex County (Connecticut) Historical Society.*

Middletown, one of the two strict Congregational Churches in the State of Connecticut founded by the converts of the great Whitfield. [George Whitfield or Whitefield (1714-1770) was a British cleric often credited with the founding of modern Evangelicalism.] The church is a bit stricter in its theology than its parent, if anything.

"But fate changed that from this respectable and orthodox church society, the blood and sinew of the greatest commercial bubble of the times should be drawn, through the masterful tactics of the preacher discovered by Mr. Ryan's committee, which was a sub-committee of 18 members that had been appointed by the church to get a pastor."

Article 57, January 12, 1950, quotes the *Boston Herald* of Wednesday, August 3, 1898:

This Rev. Mr. Jernegan is a polished gentleman. He has a classical education but he does not talk like a clergyman. He preaches a beautiful sermon and has a fine delivery, but his conversation was far from that of a clergyman, during week days, when he was on his great still hunt for a fortune in the bottom of the sea.

Rev. Mr. Jernegan's first sermons pleased because of their eloquence. In later ones it was discovered that the preacher was a little loose in his theology. He also had a way of saying things in the pulpit that left a sting. One of his earliest indiscretions was to pay his respects to the fashionable Episcopalian Society which worships in a fine brownstone edifice up the street. He said it was purse proud, that working men could not attend church there, and that other clergymen were not allowed to speak from its pulpit. He mentioned a case wherein a workingman was alleged to have been invited by an usher to leave the church on account of his dress.

This incident stirred up much feeling and reacted against Rev. Mr. Jernegan. The Episcopalians denounced the statement about the workingman as untrue. The statement about clergymen being barred from the pulpit was the result of a meeting to promote a charity work in the form of an industrial wood yard, at which Mr. Jernegan was not invited to speak, and at which the chief speaker Col. H.H. Hadley of New York spoke from the chancel steps. This was according to the custom of the Episcopal Church, the speaker being neither a preacher of the church, nor a lay leader.

Mr. Jernegan made other rash and brash remarks during the year, causing painful impressions and stirring up controversy. At the end of his first year, his salary was reduced from $1,400 to $1,200. Small wonder, then, that his mind cast about to find ways of getting more money.

JERNEGAN AT MIDDLETOWN

Article 50, November 24, 1949

While we have mentioned the wood yard venture before, the following account adds a few embellishments.

"Later Mr. Jernegan got up steam on his own account on the wood yard matter, and got tramps to work in the cellar of the parsonage sawing wood for meals. Eventually he cooperated with Hon. E. Champion Acheson, rector of the Episcopal Church, in establishing a wood yard at the foot of Court Street, near the river. Meetings were held to promote the project, one of the speakers being Rev. E.M. Poteat of New Haven who obtained some notoriety at one time by saying he 'would rather send a boy to hell than to Yale.'

"A building was bought at the wood yard, and was converted into an industrial house. The good news of the work spread all up and down the valley among the knights of the road, and a stream of hobos came into Middletown to sample the fare at the 'Tramp House,' as the Town called the place. All the 'tourists' who truck it over The Airline between New York and Boston stopped off at Middletown. The people of the city did not take kindly to this condition of affairs. They thought this particular kind of charity might be all right in a large city, where there was a resident population of outdoor poor, but it would not work in a small one. Besides, neighbors objected to the noise made at night by singing hymns at the 'tramp house.'

The merchants, therefore, declined to support it liberally. The tramps kicked, too, because the meal ticket for which they sawed wood an hour was worth but nine cents. They said that if it were worth ten cents, the price of a drink, they would work harder for it.

The *Dukes County Intelligencer* adds commentary:

> *He seemed less interested in saving souls for eternity than making life on earth more equitable…It was a period of economic misery following the Panic of 1893 and he organized a program in which homeless men could earn a small daily wage. Many townspeople were not pleased to have "hobos" wandering around Middletown.*

So the wood yard was given up, and became a commercial enterprise in the hands of Robert Pedden, a member of the Baptist Church, who later went into the employ of the Marine Salts Company at Lubec. The net result of the plan was a certain loss of cast in Middletown by Rev. Mr. Jernegan.

Mr. Jernegan had married shortly after coming to Middletown, his bride being Miss B. Evelyn Phinney, daughter of Captain Adelbert Phinney of Edgartown. There is a story in Edgartown that there was more or less romance in this marriage. When the minister brought his bride to Middletown, Deacon Ryan threw open his residence to them, and they held a reception therein, the parsonage not then being fitted up. Later the pastor and his wife had the use of Mr. Ryan's summer cottage on the shore of the Sound at Niantic. Mr. Ryan lent Mr. Jernegan money to fit up the parsonage, and was his friend and sponsor at all times. He believed in him thoroughly.

Did Jernegan Believe in Himself Thoroughly?

The *Lubec Herald* of September 24, 1933, republished an article from a Boston pastor who knew Reverend Jernegan that illuminates Prescott's perceptions two decades after the Klondike affair:

A correspondent of the Boston Herald, *Rev. Henry W. Pinkham writes to that paper that he was a classmate of the Rev. Prescott F. Jernegan of salt water gold fame and that he had a letter from Mr. Jernegan in 1927, from the latter's 26 acre ranch about 50 miles from Sacramento, Cal.*

Mr. Jernegan, who ranked as one of the brightest students at Brown University, has written several books and pamphlets, one being Man and God *which has attracted considerable comment from religious circles. Mr. Jernegan concludes his book with the following optimistic (almost fatalistic) lines;*

"I accept the Universe; it knows its business and whatever is to be, will be and will be right, for it cannot be otherwise. I am satisfied to wait and to hope. This is rather a pleasant world. In spite of terrific losses and errors I have lived long, and on the whole, gladly. I am winning to believe that I may have the indemonstrable and undeserved destiny to live again in a better world. This thin ray of hope that I shall go on building and living after death helps me. I would live well without that hope but not so well; be serene, but not happy, die resignedly but not triumphantly."

These words would appear to mark Mr. Jernegan as not wholly certain about the next world; a man possessed of truthful doubts about the future, but hopeful, believing that he will get at least more than any human being can feel that he deserves. He was undoubtedly sincere in his belief that gold could be obtained in paying quantities from the ocean, and his attitude was shared by many until it was found that it cost a dollar and ten cents to get a dollar's worth of the precious metal out of the sea. But who knows by what new methods may be found whereby Jernegan may be vindicated.

End of the Middletown Pastorate

Article 51, December 1, 1949

We have noted that at the end of the first year, Mr. Jernegan's salary as pastor of the Middletown Church was reduced as a result of various remarks of indiscretion that he made. To continue the story as recorded in the newspaper of the time, we find the following:

"By the end of the second year there were indications of an approaching crisis in the affairs of the church. The congregation was by this time divided into Jernegan and anti-Jernegan factions. Deacon Ryan stood by his protégé, and the opposition was led by Deacon Samuel C. Stevens, a mill overseer.

"The controversy soon waxed warm, and the contention finally led up to the question of Mr. Jernegan's going or staying. The anti-Jernegan faction finally decided that the shortest way to bring things to a head, inasmuch as the pastor did not feel inclined to leave gracefully, was to cut his salary some more. This was broached, but Mr. Jernegan checkmated the move for the time with a counter move worthy of the promoter of the Marine Salts Company.

"'Brothers,' he said, 'I feel that it is wrong for a minister to exact a fixed sum for his services when it is hard to raise that sum. The apostles went out without money and without price, and made their way, preaching the gospel. I am willing to do as they did.'

"The minister then said he had been doing some figuring as to amounts taken up in the weekly offerings of the congregation, had figured the expenses of the church, for fuel, lighting, and janitor service, and he was willing to take for his portion what was left after those expenses were paid out of the offerings. The Church assented, but the society held the whip hand. Contributions began to fall off. In a sermon the minister spoke sharply of those who did not do their duty by the church. This caused more hard feelings.

"Finally Mr. Jernegan said that as there seemed to be some feeling against him in the society he would tender his resignation, to test that feeling, by the vote on his resignation, he would abide.

"That was what the society was waiting for. It accepted the resignation without loss of time.

"Out of the only settled pastorate he had ever held, but still holding the friendship of A.B. Ryan and many others who believed in him, Rev. Prescott Ford Jernegan left Middletown. He sent his wife and their boy, born a year after his coming to Middletown, to his home in Edgartown, and started out as an evangelist. He was an earnest exhorter, and always had been, and his trip brought him some money. After some months, he went to DeLand, Fla. to preach, but soon came back to Edgartown, saying he was out of health."

No use in talking, he must have been a round peg in a square hole, or vice versa, and we shall see how he fared next.

Plans for Making Money

Article 52, December 8, 1949

After his experiences with the ministry, Reverend Mr. Jernegan put his energies to work in other directions. "By this time, plans for more worldly pursuits were forming in his brain. This was in the summer of 1896. He went back to DeLand, relinquished his pastorate, after having held it about six months, and returned north. At Edgartown, his old chum, Fisher, was staying, and the two were much together. Presently, Jernegan began borrowing money of friends in the town. His credit was good, owing to his father's standing and his profession. He got $500, from one Edgartown resident without security. From another, he secured a like amount in the same way. He told them the money was for the purpose of starting an industrial movement among the 'crackers' of the South. ["Cracker" was a pejorative term for poor white people.] It was to be used for buying seed for the poor crackers and farm machinery. The crops were sure to be large, and the returns in proportion, as the plan was to be a cooperative one. The trust of these friend's of Jernegan's youth must have been in reverse ratio to their information about the habits of the poor whites of the South.

"When the time came for paying this money, Jernegan was hard to find. He slipped around corners, and kept out of sight of his creditors. He said the tobacco crop was bad, and that cows had broken down fences at his industrial settlement. These stories seemed 'fishy.' He seemed to be sparring for time. Finally he paid up by check drawn on the Purity Trust Company of Boston. It happened that one of the officials of that company is a director in the local bank at Edgartown, and he revealed the fact to a few that Jernegan had a deposit of more than $3,000 with the trust company."

This was the fund with which he floated the Marine Salts Company. The crackers served their purpose well, though they were not used as the medium of getting all this money together. Some of it came from Middletown.

"While this money borrowing was going on, Jernegan and Fisher were in constant consultation, and the plan for forming the Marine Salts Company was taking shape. The scheming was done in various places. Undoubtedly, the first of it was at Edgartown.

By dint of considerable perusing of old-time papers, an interesting item comes to light. Whether it is the truth or hearsay, you may guess. From the *Boston Traveler* of June 7, 1922, is a rehash of the Jernegan story. This note concerns Charles E. Fisher. How about Fisher, his able lieutenant? The earnings of the saltwater swindle brought Fisher no more permanent gain than it brought Jernegan. After two years of wandering, Fisher died in 1900, in New South Wales, Australia. Do you believe it? Will someone please visit the cemeteries in New South Wales and find out?

The extensive archives of Martha's Vineyard Museum include a copy of the *Vineyard Gazette* from December 1902, which reports the death of Charles Fisher in Sidney, Australia. His whaling captain father, thoroughly humiliated, disowned Charles and declared him dead. Yet the museum collection of Charlie's mother Helen's correspondence holds a February 1905 letter to her from a New Zealand friend claiming that Charlie was working on a steamer headed to Australia. Captain Fisher died in November 1905. A cenotaph at his grave includes an inscription for Charles Elmer Fisher, who "died in Sydney, N.S.W." in 1902.[11]

Jernegan's Dual Role

Article 53, December 15, 1949

There is nothing more interesting in the career of Jernegan than the dexterity with which he slid out of the robes of sanctity and developed himself as a flint-hearted promoter. His clerical speech was dropped as if he had never known it. He addressed Fisher in his letters and Phelan also, as a bank breaker would address his pals. The man who was willing to emulate the apostles disappeared when Jernegan threw his strength into the effort of his life.

He impressed on Fisher and Phelan at all times that he was master in what he called "the game." He wrote Phelan, when the latter wanted to get out, that he valued his services. "The best proof of that is that you are in this thing at all," he wrote. Here was the evidence that Jernegan held his scheme then, at the beginning.

"The position I have taken on the money question" (so far as it relates to the present stage of the game), he wrote Phelan, "has received the carefulist and most protracted attention I am capable of giving it."

This was when Phelan was pressing him for pay for his part of the work. In writing to "Charlie" Fisher he urged economy, and told him to earn money for the promoting fund. "When I was in Providence (perfecting the machine), I preached to earn money. Let every man do what he can," he wrote his pal. What excellent sidelights on this complex character!

In his letter to Phelan about money, addressing Phelan as "Dear Bill," he spoke of his "reserve fund" as follows. "Such a fund is like a revolver; if you need it at all you need it very much.

Phelan insisting on having his own money, Jernegan wrote him, "I have $1,800 of debts that have cost me many a heartache and that every consideration of honor and justice should lead to pay immediately." To Fisher, he blithely wrote from the Castle Square Hotel, Boston, last February, when the "Swag" was beginning to pile up, "Let old bills go to h-ll temporarily; forget that there are lots of things we want, and act as if we had no money yet. Of course, we shall profit by our experience to date, but there is untried experience yet to come that may open up new and costly problems." [The elegant five-hundred-room hotel, built in 1894, stood at 421 Tremont near Berkeley and Chandler Streets. It was demolished in December 1932.]

In the same letter, he wrote, "There is just one time when I expect to be liberal with money, and that is the moment when, with it and us in a safe place, I sit down and divide up according to what each man has contributed to the success of the undertaking, and what he needs to fulfill his ideas of life. But until that time, I want to see every penny saved that can possibly be saved, and after that time, I swear I won't touch my principal for friend or foe…Money and lust have been the two most vexing problems of my life. I see how both may be solved if I can get a home where I can keep my body in touch with nature, and have a settled, even if small, income."

This letter was largely and widely quoted in the papers of 1898.

It might be felt that his words condemned him. "In spite of this conviction from Jernegan's own pen, there are still a few people among those who followed him in Middletown who believed he may be able to justify himself. They are not students of psychology."

"Down in Edgartown, a man who has known both Jernegan and Fisher from their boyhood summarized their qualifications for the work they carried out so successfully, by saying that 'Jernegan was the head, Fisher the hand. A better combination could not be found."

This article is copied from the *Boston Globe* of August 7, 1898. Soon, we shall get at the financial part of the story. All this descriptive matter is included that those interested in the "saga" may have all available detail. It may be some time before this little story is looked up again.

CHAPTER 8
Divulging the Deception

The Bubble Burst

Article 21, May 5, 1949

So many events of unusual interest appeared in the newspapers during the last week of July 1898—discouraging reports of the Alaskan Klondike, the "Dreyfus Affair," the last part of the Spanish-American War—that it is not surprising that there was room for yet another…rumors that all was not well with the Electrolytic Marine Salts Company.

Rumors appeared in the papers on July 28. By July 29, all the large papers carried feature articles on the nature of the company and its works. It was on Friday, July 29, that work was suspended at the "Klondike" at North Lubec.

Work had been going on steadily at Plant No. 2, with some 600 men employed in an effort to get the plant in operation at an early date. As we have seen, there were vast amounts of materials at the Canal and many enterprises connected with the construction of the plant. Of a sudden, it was announced that work would be suspended. More later about the reactions

The Lawrence Brothers Store on North Lubec Road with signs above doors. *From left to right*: DRY GOODS, POST OFFICE, GROCERIES. Note the utility poles, fire hydrant and the group of boys sitting on the porch. Situated between the Klondike factories, the store dates from 1898, according to the *Lubec Herald* magazine supplement for December 1899, and remained in business far longer. *Courtesy of Pauline Bailey.*

that this news occasioned. One thing that has been recalled by hearsay is the appalling first silence that stunned the workmen when they were first told that the work was over.

Meanwhile, ominous headlines appeared in the daily papers. From the *Hartford Courant* was this: "The Bubble Burst. Gold-from-Sea-Water Company Suspends. Steps Taken to Apprehend Jernegan." Typical of other headlines were: "Gold and Silver from Old Ocean," "Sea-Gold Stockholders. They Are Worried and Want to Know About Trips of the Officers," "The Lubec Plant Closed," "Disappointed Gold Seekers Find Seaweed," "Detectives Question Promoter's Brother," "700 Men Paid in Full," "Jernegan's Gold Mine," "Fugitives

Control Stock," "Ryan Still a Believer," "Fisher's Wedding Trip." [Ms. Bangs fails to cite sources of these assorted quotes.]

So many variations of the story of the disappearance of Fisher and Jernegan exist that it is difficult to make a definite statement. According to news copy of that week, Mr. Fisher seems to have slipped out of sight on July 20. It was given out in some quarters that he was on a wedding trip. It was thought peculiar that he should leave for such a purpose without telling his mother and sister-in-law with whom he lived. The *Hartford Times* of July 30 says in part, "Fisher's disappearance is as mysterious as Jernegan's, but it is announced that he and his wife will return next Wednesday…Fisher seems to have been Jernegan's right hand man and to have shared with the latter the largest part of the profits of the Gold secret."

No special note was taken of Jernegan's departure from North Lubec sometime after the middle of July. He had told company officials that he was going abroad in the interests of the company.

W.R. McDonald of Cromwell, Connecticut, was in North Lubec with A.N. Pierson when the crash came. He warned Mr. Pierson that Jernegan's recent behavior in New York had been peculiar. According to the *Hartford Courant* of July 29 under the caption, "Jernegan's Actions," it noted that Mr. Jernegan didn't recognize his friends; that he denied that his name was Jernegan; that he bought heavily of government bonds and that he used assumed names at a hotel and on the boat on which he sailed."

A NEWSPAPER STORY relates that on July 26, the E.M.S. Co. ceased issuing stock upon hearing that Jernegan had sailed for Europe without notice. Albert Sawyer was quoted as saying that all the stock had been sold. Stockholder Albert C. Titcomb, a former mayor of Newburyport, said, "I was sorry to hear in Boston today that the bankers of New York displayed suspicion of Mr.

Jernegan last week simply because he bought $80,000 worth of government bonds. It was with his own money, paid him by this company, and he had the right to do as he saw fit." The Honorable and Mrs. Titcomb had traveled from Newburyport to North Lubec in late June to visit the plants (see Article 19 in Chapter 5).

"Begin to Think that Things Look Bad"

Article 22, May 12, 1949

For several days after that fateful July 29, 1898, when work at the gold plant was suspended, many people, including most of the company officials, still had faith in the project. Others, including the *Hartford Courant*, "knew all along." In the August 11 issue, it was stated with smug assurance, "The collapse of P.F. Jernegan's project was sure to come. The only question was when, and the when seems to have been that point of time at which he could lay his hand on a big pile of money and slope (run away)."

In the Public Library in Boston, preserved on filmstrip, may be seen various newspapers of years ago. Commenting and speculating on the strange events of the E.M.S. Co. and its officers, the *Boston Globe* made these suggestions: "While Jernegan's actions in New York might appear strange, yet those who knew him see how characteristic his conduct in the purchase of bonds and his departure has been. That is simply his way, and he appears to have been afraid that his process would be known."

The company's president, Arthur J. Ryan, had arrived in Boston, from Lubec, on July 28. Assurances were given out to the papers under the following headlines: "Not Alarmed. Large Stockholders in Marine Salts Company Feel Safe."

"All is well," the message read. "It was known to various officials that Mr. Jernegan was going abroad for chemicals and machinery; secrecy was a requisite so no rivals could get onto his movements." It was claimed that the company would soon have machines to extract $1,000 worth of gold a day at an expense of only three or four men and a few guards to protect the mill from curiosity of people whose greed might lead them to discover the secret. It was averred that if the secret process became public, anyone having seashore property could set up a paying gold factory of his own.

It has been noted before that the main difficulties in getting this story authentic are the conflicting statements made by the press. You shall have samples of all types of editorial comment, however, and do your own thinking. From the report of a heavy stockholder–whose name was withheld–this account was designed to revive any wavering hope. "I have seen and examined the books of the Company and have seen records of bank deposits which show that the Company is solvent and can pay." That was one of the features. The company's affairs were in good order and run by men of reputable business ability.

Continuing his explanation, this stockholder said further, "Now, $1,000 a day are paid by the Company for work at North Lubec, and will continue to be until Plant No. 2 is completed. It is entirely immaterial to the Company whether Mr. Jernegan or Mr. Fisher are here or not, as they are not important factors in the success of the undertaking. It has got beyond them." Now, we might wonder who was fooling whom.

It was correctly stated that Mr. Jernegan's money was his own, paid by contract. Analyzing the situation further, the same stockholder added, "Mr. Usher, Treasurer, handles the funds. Mr. Fisher is not salaried; indeed, none of the officers, except Mr. Ryan draws a salary.

"I do not know why Mr. Jernegan has done as he did and I can only say it may be some personal freak, and genius is often-times erratic. As far as I know, he has not gone abroad to purchase any apparatus for the works, as he was not authorized to do so. It may be that he has gone over to introduce his process on the other side as he tried once before."

Always, and especially in Lubec, was the recurring thought, "Where is Mr. Fisher?" Many believed that he was hiding somewhere around Quoddy, possibly on Campobello. Others thought that he had gone to Europe. With two of the principles [*sic*] gone from North Lubec, only Mr. Pierson was left when the end came.

As late as July 28, when uneasiness manifested itself, working plans continued. News from Middletown, Connecticut, stated that "J.D. Sibley, an architect, leaves for North Lubec to superintend erections for the E.M.S. Co…He will take five men with him."

Events came to a head rapidly, and the next day found the destinies of these workmen altered. "Isaac B. Lincoln and men who went from here (Middletown) and Cromwell yesterday for North Lubec were met in Boston by Mr. Ryan who told them that owing to a change in plans, their services would not be required at present. Only Mr. Sibley went to return the tools which had been shipped ahead and to see what he could find." Indeed, what did he find?

Meanwhile, following the stunned silence of disbelief when the entire force of men was discharged, rumors and speculations arose freely. Many of the Townsmen congregated at North Lubec that night of the 29th of July in an angry and somewhat outraged mood to confer with Mr. Pierson and leading employees. Mr. F.M. Tucker recalls being among those present. The company chemist, William F. Arrington, was arrested and arraigned before Justice of the Peace B.B. Reynolds. The case was dismissed shortly for lack of evidence. [See Articles 63 and 64 in Chapter 10 for details.]

At the month's end, Mr. Jernegan was on the broad Atlantic, and there may be those who know where Mr.

Fisher was. As for the others, some sage observed that Jernegan had the cash and the stockholders had the experience, and another person pointed out that "no one can foretell the future."

It is gratifying that Mrs. Porter and Miss Mahlman have responded with such worthwhile contributions to the Klondike story. [Ms. Bertha Mahlman (1860-1929), an unmarried member of the longtime Lubec family, including Dr. Robert Mahlman, worked as an insurance agent and was also organist of the Congregational Church for many years.]

There must be many others who can send in interesting additions to this all but forgotten tale. All these things lend reality to the story of the activities of the Electrolytic Marine Salts Company.

Aftermath of the E.M.S. Co.

Article 23, May 19, 1949

It may be a surprise to some readers that Morton's Reservoir was dug by the E.M.S. Co. A pipe-line led from it to the Canal. Thus, water was to be furnished by gravity for the houses and for other purposes connected with Plant No. 2. Suppose the "Klondike" had never existed. Would there ever have been a reservoir constructed atop Morton's Hill? Aside from the practical industry of ice-harvesting, think of the pleasure this "pond" has provided for the youth, both for swimming and skating.

As stated, work was stopped on July 29, 1898. One day all was active and business-like, the next day–nothing. Work was dropped where it was, and according to one bit of testimony, "things were left in unfinished confusion."

An observation by Mr. F.S. Arnold of Middletown, Connecticut, informs us thus: "No one who has not been there can form any idea of the magnitude of the work laid out at the new plant." Among details, he mentioned the eight or ten acres visible at low tide, all "spiled" (a spile is a heavy wooden stake, similar to piling), the planking that had come and the twenty-two-foot dam that so shut in the water that none could flow out except through the accumulators.

Mr. Arnold was at North Lubec Works of the Marine Salts Company when Superintendent A.N. Pierson "made the discovery that there was no gold in the accumulators and said that Mr. Pierson's looks were enough to convince anyone that he had been duped. He was almost beside himself with grief at the discovery and immediately shut down the works and set about paying off the men. There were (about) six hundred men engaged in the works when the order came to close.

"Mr. Pierson expressed to him his belief that they had all been duped and that he was ruined by the collapse. Mr. Pierson's friends here (Middletown) are unable to account for this unless he had invested more money than his friends were aware of. The number of small investors in this city and Cromwell is large and in some cases the loss will fall very heavily, as men have mortgaged their property to invest. Mr. Ryan's friends claim that he was sincere in his belief, as he invested his boys' savings and also all of the property of his sister, who is a widow.

"There was a rumor today of a meeting about eleven o'clock Friday evening at A.B. Ryan's house in this city, the men coming in on a train and leaving later in the night, but it cannot be proven. Mrs. Ryan says her husband has not been home." Although large shipments of lumber and other supplies had been received before July 29, and although the work was left unfinished, the materials were not simply left to drift or rot away or be stolen. Provisions were made to auction or to resell the cargoes and supplies. Mr. F.S. Reynolds successfully disposed of a cargo of lumber. Mr. George Mowry bought logs that were used in the construction of a smoke house and so things went until order was restored.

"The Most Astonishing Swindle of Modern Times" did not refer to the honorable dealings of the company with the persons with whom it did business.

Exposure of the Fraud

Article 24, May 26, 1949

"North Lubec is greatly excited over the absence of the Reverend P.F. Jernegan and the consequent troubles attending the course of the E.M.S. Co. Work has stopped, pending further investigations at Boston. The 700 men have been paid in full and no serious trouble attended the 'lay-off,' but there is apparently no end to the anxious excitement which the circumstances have aroused." This appeared in the *Hartford Times* of Saturday, July 30, 1898.

There were all sorts of rumors. Here are a few comments and speculations: "There does not seem to be the fullest confidence in the Reverend Mr. Jernegan or his process for extracting gold from the sea." "The plans of selling 10,000,000 shares at $1.00 a share is a scheme for extracting gold from the pockets of generous and confiding New Englanders. It is under sharp investigation." "A gigantic swindling scheme in which Jernegan is the inventor, head manager and chief beneficiary." "A scheme of phenomenal audacity and extraordinary magnitude."

The *Boston Daily Advertiser* was one of the first papers to hazard this conclusion: "Jernegan's whole scheme is a bare-faced swindle and fraud."

People in Middletown and Cromwell were uneasy, also, but were quieted by early assurances from

Arthur J. Ryan, president of the company. Since Mr. Ryan was from Middletown, and a man of unquestioned integrity, some of the fears were allayed.

A New York investor had tried to obtain $5,000 worth of the E.M.S. stock just previous to July 28. He got only $1,000 worth and "kicked himself" that he did not get more. As rumors got around he opined, "I'm half disposed to shake hands with myself, but it may be all right yet. And if it's all right, I'm all wrong as I'm short $4,000 worth of that stock."

A comment of one of the stockholders was that "they would not view the situation with such feelings of mistrust were it not for the simultaneous absence of Charles E. Fisher, a chemist of the company who has always been closely associated with Jernegan since the process of taking the gold from the sea water was introduced. It is this phase of the matter which has led the stockholders to become uneasy."

Still, assurances continued to come from Mr. Ryan, then in Boston. Among other things, he stated that "Jernegan's departure will in no way affect the 'mining' of gold, which will be carried on as it has been, only on a much larger scale." At the same time, a prominent resident of Middletown who had left Lubec just prior to the crash was satisfied that all was well. He saw no reason why Jernegan should not buy bonds in fabulous amounts, take the trip abroad and use an alias.

Thus, the folks of Middletown received assurances that there was no occasion for alarm. They were under the impression that the work of enlarging was going on and that gold was being constantly produced. E.M.S. Co. stock was quoted at $1.40 in Boston on Tuesday, July 26, but on Thursday, after the disquieting news, it could be bought for thirty cents.

IN JULY, the newly married C.E. Fisher suddenly sailed off on a belated honeymoon. In addition to his bride, he secretly took with him about $100,000 in cash—worth perhaps $1 million today. When a dogged *New York Herald* newspaper reporter caught up with him, he claimed to be in pursuit of Fisher, whom Jernegan contended had absconded with the formula for their gold-extracting process:

> *The mysterious departure/disappearance so suddenly of the two men who had developed the process could not fail to raise suspicion. Detectives mounted an investigation, and it was not long before the sordid, embarrassing truth began to emerge. The* New York Herald *broke the story wide open with an expose in its edition of Sunday, July 31, 1898, which the* Penny Press *reported on the next day under the headline "A Gigantic Swindle."*[15]

That one man in Middletown still had faith is evidenced by the following ad in the *Penny Press* of July 28, 1898: "GOLD FROM SEA WATER/PARTIES HOLDING STOCK OF THE ELECTROLYTIC MARINE SALTS COMPANY/who wish to realize something from it, Address at once, with stamp, P.O. Box 1067, Middletown."

This appeared one day only. The man must have then altered his opinion.

This writer would pay more than the original price per share of that old-time stock. If anyone knows where a share is to be obtained, please get in touch with the *Lubec Herald*.

"THE HOLY OF HOLIES"

Article 60, February 2, 1950 (After the Bubble Burst)

Reporters for the *Boston Herald* were industrious, indeed, in procuring a firsthand account of the fall of the E.M.S. Co. On August 3, 1898, a special dispatch was sent to that paper under the caption above. Here is what the reporter found:

"Herald Man Visits the Accumulator Room. Everything Slimy, Dark, and Damp."

"Lubec, Me., Aug. 3, 1898–Today, for the first time, a newspaper reporter has been admitted to the holy of holies, the accumulator room beneath the wharf at Plant No. 1 and the secret machines that have extracted hundreds of thousands of dollars out of the credulous stockholders have been revealed to the public.

"First, you enter the laboratory, where Fisher and the others did the assaying and purifying. Then the guide, Amandas Johnson, lights a lamp, and you descend a flight of stairs, at the foot of which is a heavy plank door, securely locked. Opening this, the guide advances before you into a long room with many pilings upholding the wharf overhead and containing a flume 4 feet by 6 of heavy planks running its whole length.

"You walk along the top of this flume, which is filled with water by the automatic gates outside, and from each side of which, at right angles, many small flumes, 10 by 12 inches, lead out. Ranged along each side of these small flumes, are five or six three-inch pipes, bent so as to lead to the bottom of the open accumulators.

"These are round, kettle-shaped machines, 30 inches across, overflowing with water all the time from the pipes. Into the bottom of these kettle-shaped machines, each of which is set into a square frame of plank, were placed the mercury and acids, and the water bubbled through, floating out over the top.

"The accumulators are all connected with one another by platinum wires the thickness of a pencil. Everything is covered with a thick coating of slime, and the room is very dark. Here Fisher was accustomed to doctor the accumulators by putting into each, quicksilver saturated with gold and silver, doing the work mostly by night. He had a key to the laboratory and underground accumulator room,

Simulated accumulator in a Lubec Historical Society exhibit in a close-up showing platinum rods in an iron pot, submerged under water at high tide and exposed during low tide. *Photograph by Ronald Pesha.*

and, the night watchman says, would often work here half the night, It is a weird, uncanny place, trying to the nerves of an honest man, even in day times."

All of which makes a good story. Well, perhaps it was a bit spooky at that. From this description and a survey of the stubble of piling that remains at the Mill Pond, we might succeed in partially reconstructing that eerie set of buildings which was, at once, the old grist mill, the "Starvation" and Plant No. 1.

CHAPTER 9

Portaging Profits to Paris

Jernegan Sails to France

Article 25, June 2, 1949

On July 23, 1898, Mr. Jernegan, his wife and little boy, took passage on the steamship *Navarre* (or *La Navarre*) for France. [The 471-foot 1893 steamship was the pride of the Compagnie Générale Transatlantique, colloquially known as the "French Line."]

This followed a period of peculiar activities in New York. "The Reverend Electrolysist" sailed under the name of Louis Sinclair of 15 Market Street, Chicago, in order to protect his "secret process." One reason that he gave for so doing this was that he believed that his wicked partner, Fisher, had left for Europe on July 20, and he determined to catch him. Moreover, he was on a mission to purchase platinum and machinery for the new plant at Lubec. He said that after he purchased improved apparatus in Europe, he would return and "compel the ocean to yield up gold and silver by the ton."

Meanwhile the E.M.S. Co. had ordered Jernegan arrested or intercepted in France by an agent to "give an account of himself." On August 1, he was met on his arrival at Havre, but not arrested, as there were no legal papers. So he proceeded, with his family, to Paris.

"Jernegan in France," announced the headlines. On August 2, came the news, "Jernegan Not Caught. Eluded the Officials by Leaving His Train Before It Reached Paris."

Jernegan had left for Paris without molestation, and there were no extradition proceedings, but, as the paper said, "He seems to have deserted the train while on his way to this city." (Paris) Inquiries at depots and hotels by the mayoralty police bureau revealed no clues.

The *Boston Herald* published the following account from its correspondent in Havre on August 1: "Louis Sinclair, better known to the stockholders of the Electrolytic Marine Salts Company of Boston as the Rev. P.F. Jernegan, arrived here today on the Compagnie Transatlantique Steamship *Navarre*. The Reverend gentleman was accompanied by his wife and little boy. He traveled in an unostentatious way, occupied a cabin deluxe. He seemed disappointed that his incognito should have been penetrated. By his somewhat disconnected and decidedly contradictory remarks, it would seem that he is a much maligned man. Everyone knows that sea water contains gold and other precious metals in solution. There is no reason, then, why the Rev. P.F. Jernegan could not have discovered a way to make it disgorge its wealth.

"He is evidently a man of superior acuteness, has had a University training, and has always taken a great interest in chemistry…Strictly speaking, only half the invention was due to Jernegan's chemical knowledge. The other half was due to the electrical skill of the assistant general manager, C.E. Fisher, who, to again quote the Reverend gentleman, 'knows a lot about electricity without being exactly an electrical engineer.'

"There is, as will be seen, a remarkable similarity in the cases of these two inventors, both of whom appear to have stumbled upon a remarkable discovery in fields which they were only working in an amateur sort of way.

"The two men found that collaboration would give best results. So they joined forces. Jernegan's share of the invention was in devising of machinery that was to bring gold to the surface, so to speak, while Fisher finished the discovery by inventing a way to get the precious metal into marketable shape. Forty-five percent of the proceeds were what may be considered patent right. What the company did with the other fifty-five percent did not concern the inventors."

Next week, we shall continue the account as seen by the reporter of Havre.

As Seen by the Reporter of Havre

Article 26, June 19, 1949

"There is absolutely nothing in the nature of a fraud about our dealings with the matter," said Jernegan. "I have properly drawn up documents by which the company agrees to pay me, or rather us, 45 percent for the use of our invention. The document does not even bind me to make the workings of the invention profitable.

"They entered into the agreement with me upon the faith they had in my invention. Last Wednesday, but one, my fellow manager, Fisher, disappeared. I have reason to believe he was dissatisfied with his share of the 45 percent." The reverend gentleman did not care to tell me what that share was.

"Anyhow, he went off, carrying with him, records of his experiments. As these records are indispensable to me, I am going after them. There was no reason why I should not leave the United States. As I said, the company is entirely responsible to the public, while I have nothing further to do with it than placing my process at its disposition."

Jernegan said that at the most only a civil action could be brought against him, but upon what grounds, it is hard to say, as, from his own account, he has only used his own money.

It appears he does not know what Fisher's part in the invention was. He says he worked with Dr. Friedrich Luay at the Imperial Institute last year and found out certain chemical secrets. Then Fisher, he says, came along, asserting that he had been 'experimentalizing' in Berlin upon somewhat similar lines, but his discoveries also were incomplete. When, however, they were joined, the discoveries fitted like hand and glove, and the Electrolytic Marine Salts process was ready to be taken up by the company.

Jernegan said he had discovered indications that led him to believe that Fisher had sailed for Europe from New York on July 20. He had at once determined to follow him and hunt him up.

This was one reason why he took an assumed name, so as to catch Fisher unawares. When asked why, seeing that the bona fide character of his invention could not be called into question, he (Jernegan) had not stayed at the works in Lubec, Maine, extracting gold from his account, for the company and the 45 percent compensation, he said Fisher had taken his formulae with him, and he was tracking him to get them.

He also said he was going to Paris and thence to Berlin to see for himself if Fisher had really made the experiments he claimed.

Then the Reverend Mr. Jernegan would "probably" return to the United States. He said he should stay at the Hotel Terminus while in Paris.

To sum up: Fisher must be found, as, without him, the sea appears to be able to keep its gold.

Jernegan was asked if he did not think traveling under an assumed name gave a bad impression. "Well, you see," he replied, both the committee and everybody connected with the process from the very first kept things secret."

Thus ends the account of Jernegan's arrival at Havre. The last statement about keeping things secret from the very first was widely quoted in many newspapers of the time.

It is often asserted that if we put nothing into a matter, we can take nothing out; and as everyone knows, Fisher and Jernegan did take gold out of the saltwater.

Jernegan's Letter to Ryan

Article 27, June 16, 1949

The disappearance of Mr. Jernegan and Mr. Fisher precipitated great uneasiness among the stockholders of the E.M.S. Co. Newspaper speculations added to the tension. The people least worried, at the beginning, were the officers of the company who did not attach dire significance to the absence of Jernegan and Fisher. It may be recalled that the officers were aware that Jernegan was going on a trip, anyway. Besides, it was customary for him to be away.

In an explanatory statement of Arthur B. Ryan, president of the company, we find the following: "Of course there are many suspicious circumstances connected with the present matter. The disappearance of the men, and in fact, a few other things leads me to think that there may be something wrong, but I will not judge hastily until I hear from Mr. Jernegan."

According to intelligence in the daily papers it was stated that "today (July 30) President Ryan, of the company, expects to receive a letter from Jernegan, which was mailed in New York previous to his departure for France. The letter was addressed to Lubec, Maine, and owing to poor mailing facilities, it has been delayed. It is believed to contain a satisfactory explanation of Jernegan's disappearance."

It appears that, as Mr. Ryan was in Boston at the time, and that this letter was the key to the mystery, it was opened at North Lubec by a Mr. Bradley of Machias, acting in some official capacity. The account reads, "(The letter) was such a cobwebby affair that he (Mr. Bradley) could not make the officers of the Company here understand its contents. So he remailed it to them (in Boston) and they received it yesterday."

This letter was dated July 25 and Jernegan had sailed on the 23rd. "The most peculiar thing," sleuthed the papers, "it is that the letter is dated July 25, which is known to be two days after the Reverend Mr. Jernegan sailed for Havre. He was seen, by detectives, to say good-bye to his brother just before the French steamer sailed on the 23rd. He was also seen to land in Havre yesterday (August 1) and he could only have reached that port so soon by sailing on the steamer that sailed on the 23rd."

The presumption is, of course, that the preacher gave the letter to his brother, Marcus, to be mailed several days after it was written." Here is the letter, initials and all, just as it was written [again, Ms. Bangs cites no source for this letter or its disposition]:

"My Dear R,

"I feel something has gone wrong with F. (Fisher). He has disappeared from N.Y. after telling me that the apparatus was ruined, and has carried away the formula relating to the making of the machine in combination, which was his own invention, made since the original contract, and which I have been pressing him to make clear to me and deposit a record of. His action in this regard leads me to fear that he may have deceived me, even in the experimental stage. Since now I look back, I see he was situated where he could do so in every experiment that has been successful.

"That the main machine takes gold out of sea water is a certainty. I shall devote myself for a few weeks in trying to find him and whether I succeed or fail, I give you my word of honor that I will meet you for a conference in the near future. I advise that the directors develop under a rigid scientific test, what in our own apparatus is shown to be practical.

"Ever yours, J."

Now what can be made of that? The letter was characterized as an enigma "which might be taken for the epistle of an honest man who wanted to tell the company of the suspicious actions of his associates, or else as a clever ruse on the part of the writer to cover up his own tracks." Many interesting things have happened in Europe, of which we have no record.

ACTIVITIES IN NEW YORK

Article 28, June 23, 1949

Marcus W. Jernegan, a younger brother of Prescott Ford Jernegan, served as a secretary and helper when the latter was in New York on various occasions. There must have been many transactions that required careful planning. Near the end, Marcus Jernegan took the name of Frank W. Thompson in his dealings with certain financial houses.

Many of the deposits made by these two were in strongboxes, rather than at the teller's window. Marcus accompanied his brother to the banks in a closed cab. Only Marcus left the cab, on several occasions, leaving "our" Mr. Jernegan within to protect his "Secret Process."

The Jernegan brothers had some large dealings with certain Boston, New York and Brooklyn banks. They often dealt in government bonds on a cash basis, some purchases amounting to $75,000 at a time. All these transactions seemed on the level, and P.F. Jernegan was "well recommended" from one bank to another.

New York banks checked with those in Boston by telephone and telegraph, certifying the checks because the money was often deposited in ample amounts. However, the banks became suspicious. When Jernegan began putting so much money into government bonds, the Pinkerton Detective Agency was put on the trail of the brothers. This was on July 18, 1898, five days before Jernegan embarked for France. This investigation was started by J.E. Simmons of the Fourth National Bank of New York.

The New York assay offices stated that the operation of the E.M.S. Co. had been the subject of comment for sixteen months. It was thought remarkable that the weekly gold brick sent in from the company at North Lubec should never vary, whereas, scientists said that the amount of gold in seawater varies considerably. "We were a set of doubting Thomases, but we took their bricks, assayed them, and paid their market value just as if no one had said that they came from sea water."

ABOUT MARCUS W. JERNEGAN

Article 32, July 21, 1949

Because Marcus W. Jernegan figured slightly in the "Klondike Affair" and greatly deplored the fact that the venture was dishonest, it will be of interest to see what he did with his life.

Jernegan and others of the E.M.S. Co. and potential investors used the Lubec wharf when sailing to and from Boston, transferring to and from the ferry connecting to North Lubec. The Eastern Steamship Company's SS *Cumberland* is docked in this circa 1895 photograph. *Courtesy of Davis Pike.*

Dr. Jernegan, a specialist in American colonial history, died last February 19, 1949, at his native Edgartown. He was seventy-six years of age and active to the end.

In 1889, he graduated from Edgartown High School as valedictorian. Fifty years later, in 1939, he had as his guests at a reunion all the members of his class but one, who lived too far away. He attended Phillips Andover and Brown. From Brown University he received his MA degree in 1898. It was following this that he served as secretary to his brother.

He became principal of Edgartown High School after this. Two years later, he went to the University of Chicago, earning his PhD in 1906. From there he studied at the University of London and did research at the British Museum. Then he traveled all over the Far East, even stopping to teach school in the Philippines. It may be that his brother, Prescott was there at the time.

In 1908, Marcus W. Jernegan became a faculty member at the University of Chicago, where he remained for almost thirty years as professor of history. At the time of his retirement, twenty-one of his former students representing the faculties of eighteen colleges contributed a volume entitled *The Marcus W. Jernegan Essays in Historiography*. [The term *historiography* refers to any specific topic or body of history, such as the American colonial period. The collection of Jernegan's essays, edited by William T. Hutchinson, was published in 1937 and reprinted by Russell and Russell of New York in 1958.]

Dr. Jernegan had associates and did writing and gave lectures for a great many historical societies and colleges. Among them were the Carnegie Institute, Colonial Society of Massachusetts, the Dukes County (Massachusetts) Historical Society, American Historical Review, American Historical Society and the Quadrangle Club.

He also taught summer sessions or gave lectures at Harvard, Columbia, New York University and the University of Washington. He took part in the *University of Chicago Roundtable of Radio* broadcasts. [This radio show dealing with political and social issues was produced by the National Broadcasting Company affiliate WMAQ in Chicago. The program began in 1933 and lasted until 1955.]

Thus he became one of the first rank of American historians. The crowning work that he had planned to publish was a history of American whaling. For years he had collected notes, anecdotes and other source materials, and he was considered the foremost authority on the whaling industry. In addition, he was working on material for the *Encyclopedia Aectia* for Vilhjalmur Stefansson [1879–1962, Canadian Arctic explorer and ethnologist]. This was a study of Arctic experiences of Vineyard whale-men.

Dr. Jernegan was exceptionally well qualified for this work, being the son of Captain Jared Jernegan of Martha's Vineyard, who was a whaling master of the Arctic for forty-eight years. Incidentally, Mrs. Jernegan went on many of these expeditions.

According to the *Vineyard Gazette*, we find this opinion: "Young Marcus knew the atmosphere of whaling, the vessels at the wharves, and formed lasting impressions of the captains who were leaders in the town in that day. He went far enough to sea on a ship of his father's to know first hand some of the feel of a whaling voyage."

It is hoped that the efforts of Dr. Jernegan will be put into book form by someone qualified to do the work.

The summer estate of Dr. Jernegan was given to him by his great-grandfather on the condition that he should never follow the sea. That did not prevent him from being interested in it, however.

Another of his projects was to compile a list of logbooks of Vineyard sea captains and their present locations in various collections.

In all the excellent write-up of Dr. Jernegan's life, and that of his family, the erring brother was not mentioned at all. Yes, it certainly may be said that Marcus lived down any connection with the Gold from Seawater venture. His is a wonderful record.

Another Letter from Jernegan

Article 29, June 30, 1949

We have that Prescott Ford Jernegan's brother, Marcus, figured in the story by helping in business matters on the New York end. The following letter is self-explanatory, and you may enjoy it. From the *Hartford Courant* of September 17, 1898:

"Jernegan's Statement Made to Clear His Brother Written in Brussels.

"The Reverend P.F. Jernegan has written to the *New York Herald* a letter to clear his brother Marcus of all suspicion in the matter of the Marine Salts Company affair. He says, 'He served myself and Mr. Fisher in a purely secretarial capacity when he assisted us in New York. He was led to do this by explanations which made our actions in investing securities consistent with the welfare of the Electrolytic Marine Salts Company. If the directors of the company-men of large business experience-could receive so trustingly representations which they now repel as absurd, it is certainly not incredible that a young man just out of college should accept plausible statements from his own brother. In making a legal fight for the retention of my property, he did simply what the law gives to every accused person-gave me the benefit of the doubt. To associate his name with suggestions of fraud is as cruel a wrong as could be done to a young man whose life is spotless and who is just entering on his own career.

"Perhaps I may gain some acceptance for my words from those whose creed is now to doubt me by quoting to my own disadvantage from a letter written to me by my brother August 21, 1898 in which he says, 'Papers have been served on me as a trustee, and I have got to swear before a justice of the peace the exact amount of property of yours in my possession. This I must do for two reasons-first, because I now have serious doubt that you obtained the money either legally or morally by right, second, because I refuse absolutely to knowingly commit a perjury with self interest for all the money that ever was coined. When I went under an assumed name in New York, I did it with no selfish interest as I thought the money was morally and legally yours and that I was protecting the market value of the stock of the company. Now it is quite another matter and the truth must be told. I am disgusted with the whole business, as you can see by my last letter. However, you shall have your legal rights, though at a great sacrifice to me. It is almost incomprehensible to me, even now, how you could have the heart to carry out such a scheme. I can see how you might, so far as you personally were concerned; but here I and the family have got to go through life with this disgrace hanging over our heads at every moment. There is not a town in the land where the finger of scorn will not be pointed at us, such is the publicity of this swindle.

"'If you don't do anything else, I hope you will make some sort of a statement, stating what are the facts. You could not state anything much worse than has already been printed. Your Brother, Marcus.'

"In giving publicity to my brother's virtual condemnation of me, I do not mean to be understood as in any way admitting that it is true. My only thought in writing this is to do him a tardy justice."

Jernegan's Dealings

Article 30, July 7, 1949

When news came that the Jernegan process for extracting gold from the ocean was a fraud, his family was indignant at the idea. Had not his mother and sister in Edgartown bought stock in the company from the inventor himself? A brother in Lynn, an engineer, had also been sold E.M.S. Co. stock by his brother. And the other brother, Marcus, we have been assured did not know the true nature of the "Process," but merely served as secretary, dealing with the financial end.

Yes, the family was indignant and made statements that there was "nothing crooked" and that "everything would come out all right."

At the time of the exposure of the scheme, reporters from the city dailies fairly besieged Captain Jernegan's house at Edgartown, Martha's Vineyard. The editor of the *Vineyard Gazette*, who did not approve of this method of treating the scandal, printed only short and perfunctory reports in his paper.

It had been observed that whenever Reverend Mr. Jernegan received any considerable sum of money, he seemed to desire to travel. As stated in the papers, "He sailed for Europe (July 23,1898) with a large fortune, and his brother, Marcus W. Jernegan, who has been the reverend gentlemen's companion, in New York for the past two weeks, left for Providence, R.I. These two had New York bankers amazed at large dealings in government bonds on a cash basis, sometimes $75,000 at a time."

Of course, Mr. Jernegan did other purchasing also. To quote further, "These inventors are queer people and Mr. Jernegan has been very mysterious in his business transactions for the company. When he has purchased machinery for the plant, he has always used false names and endeavored to prevent merchants from knowing for whom or for what the machinery was to be used. He has purchased small quantities of machinery and bought it all over the country and in every way attempted to cover his transaction. He always paid cash for what he got so that the goods could not be traced.

I do not believe you could find a person who knows that he has sold anything to Mr. Jernegan and the North Lubec Works. "I can explain Mr. Jernegan's assumption of another name, or his brother using the alias of Frank Thompson in New York, only by stating that it is in line with all his actions in the past." He was "credited with a vast amount of originality."

We have seen that Marcus Jernegan did much of the business dealing with the New York banks while the inventor kept in the background. Here is an interesting bit headed, Simmon's Detectives. "Mr. Marcus W. Jernegan did not complete his last bond transaction, having apparently been frightened at the severe address of one of the Bank Presidents when he told him to withdraw his account.

"J. Edward Simmon's, President of the Fourth National Bank" and New York Clearing House, which started the investigation, has done a service to investors that cannot be too highly commended. "Others, before this, allowed the Jernegan brothers to go about their big cash transactions undisturbed and they did not seem to notice anything peculiar about their methods. When Mr. Simmons telegraphed to Boston for particulars, he got nothing but commendatory replies. He, therefore, concluded to employ detectives himself."

There are times, then, when we may all be glad that we do not have too much money to trouble us.

End of the New York Dealings

Article 31, July 14, 1949

After the searching questioning given Marcus W. Jernegan by Mr. Simmons of the Fourth National Bank of New York, no further attempts to convert large amounts of cash into government bonds were made. Directly after this, Prescott Ford Jernegan embarked for France under the alias of Louis Sinclair, while Marcus, his brother who had served as secretary and errand boy, took passage for Providence, en route to Edgartown.

Remember, that detectives had been employed to trail the brothers and when the steamer reached

Fall River in the morning, Marcus Jernegan was not among the passengers to leave the boat, as expected. There are numerous interpretations given. He was suspected, by the detectives, of trying to commit suicide but was accosted as he left his cabin by night and foiled. Another report stated that a search of the steamer was made and Mr. Jernegan was found in a private room.

The detectives engaged him in conversation, and "Marcus W. Jernegan made some damaging admissions about the methods of the company which satisfied them that the surmises of the bankers were correct and that the amount of gold to be obtained from sea water is probably not much larger than scientists had hitherto supposed."

"Where is that satchel that you brought aboard the boat?" queried a Pinkerton detective. "Oh, I expressed that to Edgartown," was the reply. "And how much money did you have in the satchel?"

It developed that there was some $40,000 or more, in cash and bonds in the valise, which the "Pinks" turned over to the Massachusetts authorities. Marcus had been instructed to send the money to his brother, in Europe, when an address was available.

In Boston, where Marcus was questioned closely, it was learned that he had been given power of attorney by his brother and what he did with the money would be legal. Marcus left Boston for Edgartown on August 1. With all that was coming to light, there is no wonder that this dismal heading appeared: "Streak of Blue Permeates the Local World of Speculation."

MARTHA'S VINEYARD MUSEUM holds collected correspondence between Marcus and Prescott. While Marcus never mentions the exact amount in the valise, he expressed "serious doubt" about his brother's honesty and integrity and later "disgust" and "disgrace."[16]

E.M.S. stock sold initially for $0.35 and reached $1.45 before the bubble burst. But conflicting data generate uncertainty as to the total capitalization of the enterprise. The Dukes County Historical Society says more than one thousand stockholders invested $750,000.[17] "It is stated that 2,428,575 shares of the stock had originally been subscribed for but that the money for only $900,000 had been paid in. About 150,000 shares were sold at 33 1/3 cents a share while the rest brought the part value of $1 a share."[18] The *Hartford Courant* said that "investors quickly snapped up $950,000 worth of the stock.[19]

How much cash did Jernegan and Fisher carry away? The Dukes County Historical Society says, "It is not known exactly how much Charlie took, estimates were $100,000. Prescott admitted that he had $150,000…and that, according to the contract he had with the company, the money was legitimately his. He and Charlie were to have 45 percent of the money raised for rights to their secret process."[20] The Lubec Historical Society book claims "at least $300,000…but more probably in the millions.[21] The *Lubec Herald* claims that "as they had access to the deposits of the company, about three hundred and sixty thousand dollars, they drew it out, Jernegan getting about one hundred and eight thousand and Fisher the rest."[22] A recent *New York Times* study finds that Fisher took about $100,000 and Jernegan "between $175,000 and $200,000."[23] The book discussing Florida investors adds that "Jernegan and Fisher had between them about $200,000, but not a dollar of it belonged to the company, according to legal agreements between the company and Jernegan and Fisher."[24]

CHAPTER 10

The Affair's Aftermath

All Quiet at Lubec

Article 61, February 9, 1950

While we have had the descriptive matter relative to the end of the E.M.S. Co., yet here is a special dispatch to the *Boston Herald*, which was published on July 31, 1898. A subtitle proclaimed:

"Workmen Not the Only Mourners. Been More a Dupe than a Partner."

"North Lubec, Maine, July 31, 1898. Everything is as quiet here tonight as could be expected with 700 men suddenly thrown out of employment, and the fond hopes of the last eight months for the building up of a prosperous city around the fabled gold producing plant of the marine salts company (In small letters, mind you) blasted beyond hope of recovery.

"The people have been for the past two days as though suddenly awakened from a dream, and they can hardly realize as yet that silence and desolation are to fill the valley which up to Friday was filled with noisy activity. The workmen are not the only mourners. They have got their pay in full, or will get it, while stockholders to the extent of many thousand dollars in the town and the city of Eastport are wondering whether or not they will get anything at all. (Guess the reporter was a little mixed there, as the stockholders were from out of state, and certainly, not many of them could have been "parking" in Lubec and Eastport.)

"A.N. Pierson, the wealthy Connecticut florist, was the only person in authority at the plant when the crash came, and while many hold him criminally responsible, others think he was more of a dupe than partner.

"Fisher and Jernegan have repeatedly claimed that only themselves and President Ryan knew the secret process. The whereabouts of Fisher and Ryan are unknown in Lubec, and no arrests have been made here. There is every evidence that the promoters of the gigantic scheme planned to carry on their undertaking for some time longer, until the big plant had been built and put into apparent operation, and many more millions gathered in for the sale of the stock of the company."

Indeed, the 'valley' is silent and desolate today. One can hardly believe it was formerly the site of so much activity. There may be those who will say that it has been quiet in Lubec ever since. Such would be deluded, of course. At least, no one in Lubec ever tried to perpetuate such a bogus industry to make for a little liveliness. Here are the rest of the *Boston Herald* reporter's gleanings: "Fisher, who was the resident manager, had often expressed his intention to remain here until the big plant was in operation. Nothing but the threatened crash, which came about for the reasons explained in the *Sunday Herald*, caused the promoters to relinquish the profitable enterprise when they did.

"Pierson was nominally the superintendent of construction, and it was under his direction that the new plant was being built. A circumstance which may or may not indicate his complicity in the scheme is the fact that all the work, or practically all of it, done in the preparation of the mysterious machine room the placing of the bogus accumulators and similar preparations, which it was not intended that the people of this vicinity should know anything about, is said to have been done by men who had been in the florist's employ for years, most of them Scandinavians who were imported here from Middletown for that purpose.

"It was claimed that the plant was located here, for the reason that no other locality in this part of the world combined the advantageous features of high tides, and sea inlets with narrow outlets, features which enable large quantities of salt water to be passed through the sluices into the machines at a slight cost. Doubtless the real reason was the remoteness of the locality from the larger cities, and the difficulty of slow transportation, which would discourage embarrassing investigations.

"It is claimed here that contractor Shanahan and bookkeeper Atwater have known that the process was a fraud, and were ill readiness to close up their part in the company's affairs when the right time came. Messrs. Jernegan and Fisher living in lavish style in houses that were built over to suit their luxurious tastes. Fisher especially presented the appearance of a man used to the best things in life, dressed in the latest style, and passed for a gentleman of wealth and education.

"Jernegan did not always have the bearing of a clergyman although the works were always shut down on Sunday, and a generally religious aspect was given to the company's affairs. Nothing was neglected to give the enterprise the appearance of genuineness. No detail of machinery of construction that would be present in a plant intended to last for a hundred years was lacking.

"The people have calmed down somewhat from the first stage of their wild excitement over the expose, and are now anxiously awaiting the action of the authorities of the law."

As we know, Fisher and Jernegan had got far away from the arms of the law. It was probably some time before either one sat down to reflect upon the rudiments of ethics.

Aftermath of the E.M.S. Company

Article 62, February 23, 1950

Here is a short account of the final shipment of gold from the "Klondike."

"Bullion From Lubec. It Reaches the Assay Office in New York, Where It Will Be Officially Tested.

"New York, Aug. 4, 1898. The assay office has received by the New York & Boston Express Company, the recorded depositor, two cones of bullion, some particles of which appear to be rich gold, and an ingot of silver with the Philadelphia Mint stamp on it. The express company's principal is the Electrolytic Marine Salts Company of Boston, and the metal came from Lubec, Maine.

"The bullion that will be assayed, weighs about 680 ounces, and if the results are the same as in the other 16 consignments from Lubec, which netted nearly $23,000, its value is about $4,000."

What a great deal of scheming it must have taken to come by that amount of the precious metal to play with, in the first place. As you see, they sent it in to the last. There was so much speculation as to the reasons the principals acted as they did. A caption in the *Boston Herald* of August 3, 1898 reads:

"Is Jernegan Innocent? Detectives Think He Is Victim of a Gang of Criminals. Believe They Have

Blackmailed and Hounded Him for Months and that He Finally Went Away, Fearing Probable Exposure. Half a Dozen Men Beside the Missing Clergyman Are Believed to Have Been in the Deal.

"The officers of the Electrolytic Marine Salts Company and the detectives employed in the cause now believe that the Rev. P.F. Jernegan has been hounded and blackmailed for months past by a gang of shrewd criminals, and finally forced to leave the country for fear of exposure.

"One of the officers of the company said last night that he was of the opinion that they could name every man of the gang who originated the swindling scheme, induced Jernegan to come in with them, to go the front and do the work, while they remained under cover and directed him in his movements. They had nothing to lose, so the detectives say, while Jernegan had everything at stake. His reputation as a minister of the gospel and his past good character were in the balance, besides the awful fear of the prison cell.

"Some of the other men, in the opinion of the police, have done 'their bit' in penal institutions, and had no fears of a new experience of like nature."

Well, that WOULD make you laugh, wouldn't it?

LUBEC LAUGHTER continues to this day. The Lubec Historical Society displays a simulated gold accumulator. A comedy, *The Down East Gold Rush* by Dorothy Blanch, was initially performed in the 1980s and again in June 2011 as part of Lubec's bicentennial celebration.

The First Arrest

Article 63, March 2, 1950

A special dispatch to the *Boston Herald*, after the suspension of activities of the E.M.S. Co., gives an account of the first arrest of anyone connected with the venture. The caption read, "William F. Arrington, Chemist for the Company, Taken into Custody and Held in Heavy Bonds.

"North Lubec, Me., Aug. 2, 1898. The first arrest at this end of the line in connection with the marine salts case was made today by Detective William Phelan of New York City, whose sensational expose of the alleged method used by Messrs. Jernegan and Fisher was published in the *Herald* last Sunday.

"The man arrested was William F. Arrington, the chemist employed by the company, who is charged with being implicated in the fraud.

Arrington was arraigned before Trial Justice B.B. Reynolds of North Lubec, waived examination and was bound over in the sum of $25,000 for appearance at the October term of the Supreme Court in Machias. He will probably be taken to the Machias jail tomorrow.

"The charges against Arrington are based on the fact that he was the chemist in charge of the laboratory at Plant No. 1 and worked in company with Manager Fisher in separating and refining the gold and silver supposed to have been taken from the accumulators in the secret room below.

"The laboratory was completely fitted out with retorts, crucibles and gas furnaces and the visitor who was fortunate enough to be admitted to this room was shown gold and silver in various stages of solution and separation, showing that actual work was done in this department of the factory from day to day.

"The *Herald* representative was shown through this room one Sunday several months ago by C.E. Fisher himself and there saw several hundred dollars worth of the precious metals, varying from what appeared to be a precipitate of silver nitrate, in several earthen receptacles, to the pure metals, ready to be molded into bricks, preparatory to shipment to Boston.

"A conversation which then took place with the much wanted manager and manipulator may have a certain bearing upon today's arrest. When the apparent risk of allowing employees to have the opportunity of discovering the secret process was commented upon, Fisher made the remark that the chemist might do all the work required of him in the laboratory, and still not necessarily know anything about the process used in the extraction of the gold and silver from the water by the patent accumulators in the machine room."

We have previously seen that Mr. Arrington was well thought of and that he was freed directly. Other details will follow. What peculiar working conditions those were for any employee. What must have been his thoughts and cogitations, if, as seems likely, he did not really know how the precious metals got into the accumulators?

Arlington Freed

Article 64, March 9, 1950

Lubec, Maine, August 4, 1898: "The Rev. Mr. Arrington was notified by the high sheriff tonight that he was at liberty, as the applicant for papers to serve on him was irregular and the evidence on which the application was made was insufficient. Thus, there were no grounds for holding him.

"Arrington has made many friends here, and is said to have $20,000 invested in the stock of the company. He comes from Newburyport, Mass., and has been employed by Messers. Jernegan and Fisher since the Plant was started. He emphatically asserts his innocence."

Other unresolved phases of the doings of this extraordinary company dealt with such thoughts that "it is the general opinion among the officers of the company that there were nearly a half a dozen men in the deal besides Jernegan. This conclusion was arrived at by the officers of the company, after they had considered many of the strange and unaccountable actions of the men suspected and many of the explanatory stories said to have been told them by Jernegan in answer to their queries about his association with the men suspected." One of the most peculiar friendships was that with Phelan, whose actions still remain a mystery.

Detective Phelan was at enmity with Mr. Arrington and stated his belief that Arrington was implicated in the fraud, and that the chemist knew the details of the fake process from the beginning. Phelan believed, however, that Supt. A.N. Pierson was innocent of any complicity in the fraud practiced by his associates.

"The detective still clings to his belief that Jernegan did not sail for Europe, despite the news that this gentleman has arrived in France. He said today (Aug. 2nd), 'I do not believe that Jernegan went to France at all. I think he is probably hiding somewhere in Maine.'" Guess Mr. Phelan did not believe even what he saw, not without reason, perhaps.

During these interesting days, so many things happened. It was reported that the dead body of an Italian was found near the works late Monday night under suspicious circumstances. "Sheriff Longfellow, who paid a visit to the Plant, and whose mission was said to be a seizure of it, says he came here only out of curiosity."

It was understood that Mr. Phelan was to remain in North Lubec several days and make other arrests, but he left on the west-bound train from St. Stephen at 7:30 August 4. Maybe it occurred to him that he might get arrested himself.

With the departure of the people connected with the Klondike affair, Lubec had only to settle back and in time almost completely forget that such an event ever took place.

About Stetson K. Ryan

Article 33, July 28, 1949

As a matter of interest, let us follow the life of another one or more of the people affected by the Electrolytic Marine Salts Company. We recall that Arthur B. Ryan of Middletown, Connecticut, was the president of the company. He was the jeweler who believed wholeheartedly that the metal that came from the Jernegan accumulators was gold. He was right, of course. After the "secret process" was learned to be a hoax, Mr. Ryan's life was affected unpleasantly. He did not reenter the jewelry profession. He did work at several other occupations until his death.

Mr. Ryan had three sons, who as young boys were in this town that summer of 1898. The subsequent reactions of the public concerning this unusual company caused many a heartache. Part of the anguish might be summed up in these words of one who followed the story: "Everybody laughed." But as one son, Arthur B. Ryan Jr., said, "That is water over the darn." He claimed never to have read any of the details in the papers of the times, although he works for the *Middletown Press*.

Another son of Mr. Ryan is Leonard Ryan, judge of probate and exceptionally well liked. One hears of many kind things that he has done. All are active in the Baptist Church where Prescott Jernegan preached in 1892–95.

A third son is Stetson K. Ryan, who wrote a history of the church mentioned above. Of Pastor Jernegan, he states, "Later he went to the Philippine Islands to engage in educational work under the sponsorship of the Federal Government. His latest book published is *Man and His God*. He is now residing at Mill Valley, California, and receiving a modest government pension."

This was written in 1945. The Ryans receive Christmas cards from Reverend Jernegan as a rule. A letter to the Mill Valley address was returned unclaimed, to the writer, last winter. It is averred by some, however, that our Mr. Jernegan is still living. It remains to be seen.

To get back to Mr. Ryan, he has recently received signal honors 'for outstanding work with the blind." Following are quotations.

"Stetson K. Ryan of Middletown, Connecticut, is one of its public servants of whom Connecticut hears very little. However, he well deserves the honor which was paid him yesterday in New York when he received, from Helen Keller, the Migel Medal for 'marked contribution' in work for the blind." [According to the American Foundation for the Blind website, Ryan received this award in 1949.] "What is it to Stetson K. Ryan, the jeweler's son and former newspaperman? 'My work with the blind,' Mr. Ryan says, 'is my life.'"

Mr. Ryan has been executive secretary of the State Board of Education for the Blind for more than 25 years. [According to the Connecticut State Library's webpage of agencies for the blind, Ryan served as secretary from 1918 to 1933 and executive secretary from 1934 to 1954.] The duties of the board are to advance the opportunities of the sightless to obtain education. However, Mr. Ryan has done much more than that. In the many years he has held the position, he has served the blind people of the state as their official friend and adviser. Of some 2,300

Fisher's house on North Lubec Road. In this house, Charlie Fisher secretly cast ingots from old jewelry, as long as his supply lasted. The residence still stands, north of the old Lawrence Store. *Photograph by Ronald Pesha.*

in Connecticut over twenty years old, hundreds have jobs due to Mr. Ryan's efforts.

Of the homes occupied by top officers of the E.M.S. Company, all burned in the August 6–7, 1921 North Lubec fire except Charlie Fisher's, which was near the Lawrence Sardine factory. Within this house, Fisher melted down and otherwise processed whatever gold he could acquire to cast ingots for the picaresque plot.

The Secret Closet

Article 20, April 28, 1949

"People always said that there was a secret closet somewhere in the house and that the gold was hidden in it." This comment was made by Mrs. George

Comstock in conversation with Lloyd D. Nugent in 1915. The house referred to is now lived in by David Porter and family. During the Klondike affair, it was occupied by Charles E. Fisher, assistant manager of the E.M.S. Co.

Mr. Fisher it was who engineered the putting of the gold into the accumulators, which were beneath the wharf of the old Klondike, or Plant No. 1. When the darkness and the tide served right, Mr. Fisher, aided by his trusty diving suit, would "salt" the boxes with gold. It must have taken careful planning to approximate the correct amounts each time, as well as to deposit it unnoticed. Not only that, but the obtaining of the gold must have been one long source of worry. The people who actually bought up the gold in the large cities must have worked carefully to avoid suspicion.

Let Mrs. Nugent continue: "It was the first winter that Lloyd and I stayed with his folks at that house. Well, we didn't say anything about it to anyone, after Min told Lloyd, but one stormy Sunday we put on some warm coats and went up into the attic. We couldn't think of anywhere else where the closet might be. "The attic was beautifully finished with varnished wainscoting. We kept tapping and pressing the wainscoting up one side and down the other and finally we came to the wall behind the closet door. The noise from our thumping sounded different there, and a little pressure made one of the panels give, so I pressed a little harder and pushed toward the left and the panel slid out and showed a shelf.

"Right then I couldn't see anything, but Lloyd said, 'Reach way in. Maybe there is something there.' I did, and got hold of paper and pulled it to the opening and found I had one of the old-fashioned striped candy bags in my hand, and that there was something in it. Lloyd said, 'Some of old Fisher's gold, I'll bet.' Sure enough it was, five or six lumps. We never told anyone that we did find it. We slid the panel back into place and took the bag downstairs and tucked it away in a bureau drawer."

Sometime after that Mr. and Mrs. Nugent moved, and the striped bag was overlooked, or forgotten. Later, when they tried to find it, it had disappeared. It is thought that it was thrown out, accidentally, as being of no account. So even now, the small lumps of gold may be down over the bank somewhere near. Four shelves, small and neat, now show where several pieces of the wainscoting have been cut out. It makes the dearest little closet you ever saw. Wouldn't it be interesting to know how much gold was concealed there a half century ago?

Relative to this, Mrs. H.P. Farnsworth once had a vial containing some of the famous gold in her possession.

After the E.M.S. Co. settled up its business in North Lubec, the Fisher house was put into the possession of Rufus Kilgore of Melrose, Massachusetts, who had invested heavily in the company's stock. The house was bought by I.M. Bangs from Mr. Kilgore, and the deed is still in existence.

What Happened to Jernegan?

"In late July 1898, a former friend of Fisher's [presumably, the detective William Phelan] who knew about the plan from its earliest stages blackmailed Jernegan and Fisher. He demanded that they either pay him, or he would leak their secret to the press.

"Fisher fled to New Zealand with $100,000, and died a few years later in Australia. [For more about Fisher's uncertain final days, see "Plans for Making Money," Article 52 in Chapter 7.] Jernegan set sail for Europe with his wife, son, and $150,000. After a year of guilt bearing down on his conscience, Jernegan returned $144,000 to the treasurer of the company and further authorized him to withdraw another $35,000 from Jernegan's bank account in the States. After his wife left to go back to the U.S. with her son, she was granted a divorce on grounds of abandonment.[25]

"In the October 1898 *New York Times* Jernegan's father announced that he had received a letter from his son writing from Bursells, saying that he had made up his mind to let the law take its course. The former minister contemplated returning to the United States to surrender himself and return property to the sea gold company's directors. He planned to hand over his cash, books, and bonds and other securities, and if liable, stand trial for his actions and suffer punishment."[26] However neither Jernegan nor Fisher ever returned.

"He was very sorry for what he had done and offered to pay back his creditors but did not do so. He spent tremendous sums in France. The American government could not extradite him. He gambled in every haphazard venture then in vogue and rumor has it that he lost vast sums in a 'gold from sea water' scheme which was started in England.

"In 1902 he was in Washington but the news leaked out and he went to Manitoba. In 1903 he was discovered as a teacher in the Manila Normal School in the Philippines. He was penniless, and after being discovered he disappeared."[27]

ACCORDING TO THE ARCHIVES in Martha's Vineyard, Jernegan went to Alaska to pan Nome sands for gold. Netting only $200, Prescott sought, and found, a career teaching in the Philippines. Between 1901 and 1910, he wrote *A Short History of the Philippines* for use in the schools. Crawford cites *Who Was Who,* Vol. 4, as claiming that Jernegan wrote the national song "Philippines My Philippines," but the website of the Filipino Music Collection of the Filipinas Heritage Library names the composer as Francisco Santiago. Jernegan apparently collaborated.

Returning briefly to California, he then steamed to Hilo, Hawaii, as a teacher from 1911 to 1921 and then became principal at McKinley High School in Honolulu until 1924. Apparently, he retired to California. Jernegan died on February 23, 1942, during a visit to Texas.

Gold from Sea Tides

Hartford Courant, *January 17, 1926 (used by permission)*

How different the whole aspect of the present-day attempt! The man at the head of it today is a man of science and has seen no vision, save the vision of prosperity for Maine and the industrial revival of New England. The process is no secret. There is nothing new about it. The same principles are applied here that have been applied since the earliest man made a wheel turn around by the force of falling water. There is no belief necessary. As the scientist himself said, "One needs only to turn a glass of water upside down to discover that it will fall'"

Dexter P. Cooper, who thought out the plan and who will put it into operation when the time comes is not without experience in engineering. In the world of science he is known as the man who electrified the Milwaukee and St. Paul Railroad throughout the state of Montana. He is the man that dammed the Mississippi River at Keokuk, Iowa, and he is the man who built the enormous dam in Chile for the Guggenheim interests there. [Guggenheim mining interests profited from major deposits of Chilean nitrates, essential for explosives.]

The great power plant as Muscle Shoals was built by Mr. Cooper's brother, Colonel Hugh Cooper, and Mr. Cooper himself was on the engineering staff. [Woodrow] Wilson Dam, which has made Muscle Shoals a synonym for the generation of enormous electrical power, is partly the handiwork of Mr. Cooper, and as the Muscle Shoals project is directly in the line of his present work, his experience there is a guarantee of the success of the present venture. [Construction of Wilson Dam on the Tennessee River near Muscle Shoals in northwest Alabama began in 1918. The impounded water generated power for plants extracting nitrate for use in munitions.]

It was by the merest chance that Mr. Cooper became interested in the tidewater scheme in Maine. Instead of sending a vision to Mr. Cooper while he lay in a sleeping car, heaven sent appendicitis, which laid him up in his summer place in Campobello for over a year. While lying out in his invalid's chair, he watched the tides and, when well enough to get about, started his survey of the land formations thereabout. His staff consisted, not of scientific men, but of fishermen who had spent all their lives in the two bays, and with their assistance he plotted out all the currents, depths, tides and other necessary information for the great experiment…

What will this mean? Cheap power for New England; New England in the same admirable position she held before the introduction of steam and the importation of coal from the Pennsylvania fields. A revival of manufacturing in the East, with New England holding a commercial position that cannot be threatened for a century. This and more is included in the vision that came to Mr. Cooper when he lay in his easy chair on his summer porch in Eastport, Maine.

But it is not in the vision, nor in the way the vision came that differentiates him from thousands of other men throughout the last sixteen hundred years who thought that they too saw the handwriting on the wall and set forth to make some king or commonwealth rich by their methods. The extraction of wealth in the form of gold from other elements by no means started with Reverend P.F. Jernegan, nor with Mr. Cooper.

Science is the modern wonder worker. But science in its present form owes much to those men, the alchemists, whose direct descendant Mr. Jernegan is. Alchemy has fallen upon evil days; its ancient heritage and birthright have been forfeited because of the ill fame its later practitioners brought upon it…

That interest in peculiar and cryptic things was rampant at the time is manifest by the fact that the financial clergyman took a large amount of gold to France. Few doubted that he was bluffing, but it took a tremendous number of chips to call him. Then instead of laying down his hand, he took it with him.

Mr. Cooper stands with both feet firmly imbedded in the twentieth-century religion, science. Mr. Jernegan is the posthumous son of a decadent alchemy for which a belated lament had arisen, permitting his hoax to catch the public fancy long enough to net the inventor nearly a million.

What Mr. Jernegan added to the processes of the alchemists was modern business methods, the office, the secretary, the executive meeting behind closed doors where little more went on than is customary at most such sessions. Being the perfect showman, Mr. Jernegan offered the various elements in the public whatever they called for. To the occult minded he gave the mystery of the alchemist, gold from seawater whence it had never come before. To the businessman he offered large-scale production and all the trappings and fanfare of commerce in millions. That was his contribution. Jernegan–the ecclesiastical Barnum–the too practical visionary–the last of the alchemists.[28]

Long before he became president, Franklin Roosevelt also witnessed from his Campobello compound gold in the Fundy Bay tides: cheap power from the sea. The Quoddy Project, headquartered on Moose Island west of Eastport in 1935, began building Dexter's dream: dikes and dams driving dynamos. But before completely cutting off Cobscook Bay waters, the funding, well, dried up, for the Great Depression lingered long.

But 2012 brought sophisticated twenty-first-century technology to the Bay of Fundy. On August 10, 2012, the Ocean Renewable Power Company of Portland, Maine, submerged the first of twenty underwater turbines to the sea floor near Eastport, Maine. Powered by Fundy's tremendous tides, the turbines extract power from the sea hardly five miles from Jernegan's hoax as the golden eagle flew before becoming extinct in Maine. And unlike the past, studies address environmental factors.

Concluding Thoughts

Gold as metaphor for abundant wealth permanently adheres to our language in the technologies that have wrought riches undreamed of in 1898. But an inexpensive way to accumulate *metallic* gold technology has yet to discover. If it comes to pass, and gold without limit gluts the globe, would its worth not dwindle? Consider a 1961 episode of the classic early television show *The Twilight Zone*. "The Rip Van Winkle Caper" concluded in the year 2061, when industry had learned to manufacture gold cheaply and it became worthless. In suspended animation for a century, the thieves awoke and tried to buy their way with valueless ingots.

Increasingly sophisticated attempts to extract gold profitably by individuals from eccentrics to scientists continued through the twentieth century to this day. German chemist Fritz Haber (1868–1934) sought a successful process that would pay his nation's debt after World War I. The 1919 Treaty of Versailles levied reparations on Germany totaling 132 billion gold marks. Haber had earned a 1918 Nobel Prize for developing inexpensive ammonia (valuable for fertilizer and explosives), and he is sometimes called the father of chemical warfare for developing and promoting the use of poison gas during that war to end all wars. But he failed to develop affordable gold extraction. Analyzing .004 parts per billion of gold in seawater, he concluded no known process could be cost-effective.

If you expected a do-it-yourself hint from this book, consider this: A milligram of gold resides, dissolved, in a ton or 233.86 gallons of seawater. Working the math, to extract one ounce of gold requires processing 662,300 gallons of ocean at 100 percent efficiency. For comparison, an Olympic-sized swimming pool is 660,430 gallons.

No chemical or industrial process is 100 percent efficient. You're lucky if it's 10 percent. By the way, a gold medal earned in an Olympic pool contains 92½ percent silver and about 1⅓ percent of actual gold.

And not from seawater.

Addendum

Lubec First Down-East Town to Be Manufacturing Center

Jeremiah Fowler's Company at North Lubec Harnessed Tides at Canal to Furnish Power for Lime Plant and Made Lubec "Boom Town" of Early Days

By F.W. Keene, editor, Lubec Herald, *November 21, 1941*

The resourcefulness of our forefathers is proverbial. Placed here on rugged soil, and surrounded by dense forests and rushing waters, they were obliged to make use of whatever was at hand to gain a living.

Here in this far eastern section, with farming, fishing and lumbering being carried on, they managed somehow to live comfortably, although there was a lack of manufacturing industries for want of power. Steam power had not reached so far east as to be employed here; electricity would not be heard of until the first incandescent light had been turned on at the Centennial in Philadelphia in the summer of 1876; internal combustion engines were still far in the future. The sole answer to the question of power was water. On the St. Croix, there were mills which sawed long and short lumber, but on the Lubec promontory there were no rivers.

Knew Tide's Power

But they knew of the vast power of the tides as many other Maine communities did, although particularly favored by the great inrush of tides into the bottleneck of the Bay of Fundy and thence into the estuaries of Passamaquoddy. This natural resource had already been employed farther up the Cobscook at the mill of Mr. Bell in Edmunds, still referred to as the "Tide Mill." [Robert Bell of Scotland settled this land in 1865; it still remains in the family as Tide Mill Enterprises and Tide Mill Organic Farm.]

How could this power be used to provide employment for those who did not farm and could not fish offshore because of the fickleness of the weather in winter? True, there were a few small plants where herring were pressed for the then valuable oil, and the residue used to grow crops. Charles Treat had such a plant on Allan's Island, to which the first railroad in this county led, running from his weirs along the shore and bearing cars filled with gleaming herring for the "press kettle." Daniel Ramsdell had come from Nova Scotia with a process for smoked herring, but the demand was limited to the output of a few weirs and nets.

Early Lobster Packer

Mitchell, of Scotland, had landed on this same island. which has passed through many radical, changes and was very secretly putting lobsters in tin cans and sealing them with a copper [soldering "iron"]. Many Peeping Toms were trying to discover and imitate the process, for lobsters were under every wisp of seaweed, and labor was cheap.

But Lubec wanted a real year-round industry, and the people of North Lubec, headed by Jeremiah Fowler, finally hit upon something which caused their section to enjoy what in this day would be called a boom.

The tides run here with great force, up to six miles of the neck shore, around Gove's Point, and eddy back into South Bay, where the elevation is higher than that of the foreshore, making a natural reservoir in extent, though rather shallow. Why not let this back current flow through the land at some narrow point, and employ it to turn the wheels of industry?

Tides Harnessed

From Machias was brought a surveyor named Jones, and in company with another of the same profession, Jonathan Weston, lines were run across the "waist" of the promontory, through a high and rugged cliff, for a distance of not more than 2,000 feet, connecting with South Bay. Against the western end of this they found the tide rising 12 to 15 feet twice a day.

With a few feet of this barrier as a coffer dam, they set about building a high dam of rock and timber, put in sea gates and then at the formal opening, which was a gala day at the "canal," blew out the earth and ledge, waited for high water and with a speech by Mr. Fowler and another by Mr. Comstock, the gate was lifted and the torrent of boiling white salt water roared down the declivity, under the wooden traffic bridge and into Rumery's Bay. It appeared to furnish enough power for some industry. Obviously the inflow must be held and used when the tide of the bay had gone down enough to allow an adequate fall.

A little before that, lime had been discovered on the neck, and the residents were accustomed to quarry it out and burn it for their own use in plastering houses, when mixed with the hair of slaughtered animals. Why not dig and grind and sell it to others? They did not, however, put all their eggs in one basket. If the lime played out, they would build a lumber mill and cut off the heavy timber from the tip of the neck to the mail road to Machias, or perhaps put in a hopper and induce the people to raise corn and oats.

Aid from Washington

The lime did play out, long before one stick was put upon another for a mill, but they had the power and did not intend it to be wasted. The story of their efforts had somehow got into the columns of the old *Eastport Sentinel*, and finally to Washington, where it was read by the representatives in Congress from the Passamaquoddy, Josiah Bartlett and Moses Fuller, who had been the town postmaster.

Meantime, Mr. Fowler had received a letter from a cotton planter in South Carolina, asking if his mill was operating, and if he could furnish him with a few thousand barrels of lime rock plaster to be combined with material from the fossil beds of that state as a fertilizer for the cotton crop. Mr. Fowler called a meeting of the backers of the project, who had put considerable effort and money into it. There was one way out.

Over at Hillsboro and Parrsboro, in New Brunswick the huge deposits of lime rock, white and blue, were already being manufactured, but a high duty kept the manufactured article out of the United States market. If the raw rock could be imported in schooners, at a not too great cost, it might be the answer.

Two delegates were appointed, and set sail for New York on board the *Three Brothers*, which was loaded with salt fish, potatoes and smoked herring and from there went overland by coach to the capital where they conferred with Representatives Bartlett and Fuller.

TIDE MILL BUILT

The arrangements made were very satisfactory. Plaster could come in raw at a nominal duty, and up went the mill, amid great rejoicing. The equipment included a big undershot wheel with buckets, so as to get the fall until the tide flowed up under the tail race, a crusher with wide iron plates, one stationary and one working on an eccentric shaft, a V-shaped "cracker" which sat under and received from the crusher the coarse plaster; a hopper and endless belt which took the pulverized plaster up to the storage bins, two stories high. From here it fell by gravity down a chute and into bags and barrels.

Up sprang a cooper shop and a stave mill, a heading saw was belted on to power the shaft. "Hoop-shavers" went into the small hardwoods on the Neck, cut and bunched hoop material of white and yellow birch sapling, brought them in and shaved them to the required thinness for encircling the barrels. Work was made for all young and old.

There was no depth of water at the mill for docking vessels, but on the north side of the "Point" there was ample depth. Here went up a series of long wharves, connected by a causeway. On these wharves was a custom house, with Stephen Thatcher in charge. Two- and three-masted schooners from up the bay landed their dusty cargo, paid the duty and sailed away for more. After the second trip back, they found cargoes of ground plaster awaiting them, to be carried to New York or Philadelphia or, in some cases, still farther south for reshipment into the cotton belt.

Business was booming here. New stores were opened on the waterfront. The old taverns, the Golden Ball and the Chaloner, were filled with roistering sailors. Men came on foot or by stage, put up and looked for a berth.

Scows and lighters piled between the "Point" and the mill. Wages were a dollar a day. New homes went up, a public hall was erected and another schoolhouse, for families were moving into town.

BLAZE ENDS BOOM

And then, one night when there was a "spring" tide, came the end of the great doings at the Canal. The square plank penstock was full, the pressure was spinning the main shaft merrily, but it was too near a partition, and a "hot box" set a fire which roared to the top of the elevator shaft and destroyed the entire plant.

There was weeping and wailing, for work was ended, and folks started moving away. The loss was great, and the mill was not rebuilt. But a concern up the St. Croix [River], seeing the possibilities, built a large plant at Red Beach, and many of the North Lubec people moved there and went on as overseers of a new crew. Red Beach took over the industry, and not only that, but imported the South Carolina fossil deposits, and ground them with the plaster and with bones, shipping the completed fertilizer south.

Today, the Red Beach plant is gone. ["Fire of unknown origin totally destroyed the Plaster Mills at Red Beach Monday, with all adjoining buildings and the wooden highway bridge across the Mill Stream," *Lubec Herald*, September 2, 1926], the South has turned to other, simpler and less bulky cotton fertilizer, and the art of grinding raw plaster is forgotten in the first Passamaquoddy town to break away from primitive pursuits and undertake to become a manufacturing center.

Later on, the Canal was used in the preliminary experiments of Jernegan, who instead of turning wheels with the impounded water from South Bay, ran it through "accumulators" to extract the gold, but succeeding only in attracting and accumulating the cash of investors in Massachusetts and Connecticut.

Today the Canal is spanned by a 20-foot high solid traffic fill and bridge. The tide runs up to it on both sides, and above the bridge small boats come and go with wood and clams and periwinkles. [The old Plaster Mill Road is marked with a sign on the west side of North Lubec Road just north of the Canal Bridge.]

The old-timers are all gone. Nobody, remembers the mill: the promoters are long since forgotten, but some of the middle-aged ones remember what their grandfathers told them about the great days when the town turned out en masse to see the lever thrown that started the first piece of plaster through the crusher in Mr. Fowler's mill.

Notes

Chapter 1

1. Chapman, "Great Lubec Gold Swindle," 85.
2. Railton, "Jared Jernegan's Second Family," 74.
3. Crawford, *Florida's Big Dig*.
4. Ibid., 106–08.
5. *Hartford Courant*, "Dredging Gold from Seawater," January 17, 1926.
6. McCain, "Get-Rich-Quick Scheme."
7. Railton, "Jared Jernegan's Second Family," 78.
8. Murphy, "Gold from Seawater Swindle."

Chapter 2

9. Murphy, "Gold from Seawater Swindle."
10. Railton, "Jared Jernegan's Second Family," 92.
11. Avery, handwritten note.

Chapter 5

12. Chapman, "Great Lubec Gold Swindle," 88.

Chapter 6

13. Railton, "Jared Jernegan's Second Family," 92.

Chapter 7

14. Railton, "Jared Jernegan's Second Family," 93–94.

Chapter 8

15. McCain, "Get-Rich-Quick Scheme."

Chapter 9

16. Railton, "Jared Jernegan's Second Family," 88, 90.
17. Ibid., 80.
18. *Portland Sunday Telegram and Sunday Press Herald*, 1941.
19. McCain, "Get-Rich-Quick Scheme."
20. Railton, "Jared Jernegan's Second Family," 80.
21. Winston, "Great Lubec Gold Swindle," 85.
22 Aye, "Bill Aye Recalls."
23. McCain, "Get-Rich-Quick Scheme."
24. Crawford, *Florida's Big Dig*, 112.

Chapter 10

25. Murphy, "Gold from Seawater Swindle."
26. Crawford, *Florida's Big Dig*, 114.
27. *Hartford Courant*, 1926.
28. Ibid., January 17, 1926.

Bibliography

Avery, Myron. Handwritten note, 1926 or shortly thereafter.

Aye, Bill. "Bill Aye Recalls Klondike Gold Extraction Scheme." *Lubec Herald*, April 4, 1946.

———. "'Klondike' Folded Up When Investigators Found Diving Suit." *Lubec Herald*, April 18, 1946.

Bangs, Carrie. Weekly series. *Lubec Herald*, 1949–51.

Blanch, Dorothy. *The Down East Gold Rush.* Unpublished play by Lubec native Dorothy Blanch (1919–2008), Lubec native and professional actress. Circa 1980.

Chapman, Winston "Dick." 1976. "The Great Lubec Gold Swindle." In *200 Years of Lubec History, 1776–1976.* Lubec, ME: Lubec Historical Society, 1976.

Comstock, George Hiram. "Old Tide Mill at North Lubec Has an Interesting History." *Lubec Herald* July 21, 1927.

Crawford, William G., Jr. *Florida's Big Dig: The Atlantic Intracoastal Waterway from Jacksonville to Miami.* N.p.: Florida Historical Society Press, 2006.

Hartford Courant. "Dredging Gold from Seawater." Sunday, January 17, 1926. [The full front page of Section 5.]

Keene, Fred W. "Lubec First Down-East Town to Be Manufacturing Center." *Lubec Herald*, November 21, 1941.

Portland Sunday Telegram and Sunday Press Herald, December 7, 1941.

Railton, Arthur R. "Jared Jernegan's Second Family." *Dukes County Intelligencer* 28, no. 2 (November 1986). [Published by what is now the Martha's Vineyard Museum, its archives appear to be the greatest treasure-trove of documents relating to Prescott Jernegan, including his unpublished 1934 autobiographical manuscript.]

200 Years of Lubec History, 1776–1976. Lubec, ME: Lubec Historical Society, 1976. Reprint with addenda, 2010.

Winston, Chapman. "The Great Lubec Gold Swindle." In *200 Years of Lubec History, 1776–1976.* Lubec, ME: Lubec Historical Society, 1976.

Internet Sources

Brodrick, Sean. "Maine's Gold Rush in the Sea." Red-Hot Resources. http://redhotresources.blogspot.com/2007/07/maines-gold-rush-in-sea_03.html.

Cherry Creek News. http://www.thecherrycreeknews.com/businesslbrgopportunities-othermenu-66/141-investor-info/1478-the-gold-rush-in-the-sea.html.

"Girl on a Whaleship," fine reading about the childhood of Prescott Jernegan's sister. http://www.girlonawhaleship.org/newspapers?nid=2457&dat=19871117&id=4q5JAAAAIBAJ&sjid=sw4NAAAAIBAJ&pg=5840,4983730.

Graettinger, Diana. "Lubec Firm Promised Gold and Silver from Seawater in 1898." *Bangor Daily News*, November 17, 1987. http://news.google.com.

Market Oracle. http://www.marketoracle.co.uk/Article1501.html.

McCain, Diana Ross. "Get-Rich-Quick Scheme Turned into a Scam." *Hartford Courant*, September 16, 1998. Available online at http://articles.courant.com/1998-09-16/news/9809130034_1_tests-sea-funded.

Murphy, Jackson. "Gold-from-Seawater Swindle." NYTimes.Com. http://www.mvtimes.com/2012/07/25/prescott-jernegan-gold-from-seawater-swindle-11663.

Museum of Hoaxes. http://www.museumofhoaxes.com/hoax/archive/permalink/the_gold_accumulator.

NYTimes.com. August 1 and July 27, 1898. http://query.nytimes.com/mem/archive-free/pdf?res=9800EED71738E433A25752C0A96E9C94699ED7CF; http://query.nytimes.com/mem/archive-free/pdf?res=F40D14F63C5C11738DDDAE0A94DF405B8885F0D3.

Old News from Southern Maine. http://www.someoldnews.com/?tag=gold. [Describes other "gold rushes" in Maine.]

Panning for Gold at Ogunquit's Perkins Cove in March 1860. Several of the capitalists who witnessed Jernegan's successful demonstration were summer residents of Kennebunkport. Rather than purchasing expensive shares of his company, they made plans to reproduce his apparatus on Fort Island. May B. Whiting, regular contributor to Henry Ford's publication, the *Dearborn Independent*, wrote about the Stage Harbor Gold Rush in 1926. "On Fort Island (Tiny Fort Island, connected at low water to Stage Island), off Cape Porpoise (a small coastal village in the Town of Kennebunkport), they erected a clubhouse but instead of the customary furnishings they installed pumps of the largest and most expensive make. For a week they pumped. They pumped the ocean side and they pumped the harbor side, but mud remained mud and sand remained sand with never a gleam."

Melville Freeman wrote, "The little building which now stands on the island was built around the turn of the century to house the large water tanks belonging to a corporation which was formed to extract gold from sea water. It was found that gold could indeed be obtained in this manner but at a cost of about five dollars for every dollar's worth of the precious metal." In the August 1898 *Wave*, it was reported that the furnished cabin at Fort Island was a perfect picnic retreat. Clearly, all gold prospecting had ceased.

"The Rise of the Electrolytic Marine Salts Company." *Market Oracle*. http://www.marketoracle.co.uk/Article1501.html.

Spilman, Rick. "Lubec's Gold from Seawater Hoax." The Old Salt Blog. http://www.oldsaltblog.com/2010/04/01/lubec%E2%80%99s-gold-from-sea-water-hoax.

"The Ten Greatest Swindles in United States History." Listosaur.com. http://listosaur.com/history/10-greatest-swindles-in-united-states-history.html. [Jernegan the tenth, Bernie Madoff the first.]

Index

About the Author

Carrie Bangs (December 3, 1898–October 17, 1975) lived and wrote the primary text of this book in the house (no longer extant) once owned by Electrolytic Marine Salts Company president A.B. Ryan. Never married, she was daughter of Albert Leon (1865–1946) and Mary E. Comstock (1869–1946) Bangs. All are buried at Major Lemuel Trescott Cemetery, North Lubec. According to Edith Comstock, native and lifelong resident of North Lubec, Carrie had a teaching career at an all-boys' school in Hartford, Connecticut, returning home every summer. Another native North Lubecker, Peggy Bailey, said that "she retired from teaching and came home permanently in the early sixties. She was a dedicated volunteer at the library, and spent much time cataloguing the collection."

Carrie Comstock Bangs (1898–1975). *Courtesy of the Lubec Memorial Library.*

The Major Trescott Cemetery on North Lubec Road includes the remains of Albert Leon Bangs (1865–1946), Mary E. Comstock Bangs (1869–1942) and their daughter Carrie Comstock Bangs (1898–1975). After serving in the Continental army during the War for Independence, Major Lemuel Trescott (1751–1828) settled in Down East Maine. When the British occupied Eastport in 1811, he moved to North Lubec, occupying property at Bangs Point acquired by his first wife, sister of James Avery, who was granted a lot as a Continental army veteran. *Photograph by Ronald Pesha.*

About the Editor

Ronald Pesha retired to Lubec with his wife, Ronna, in 1988 after dual careers, seventeen years in commercial radio and television followed by twenty-two years teaching at SUNY-Adirondack in Queensbury, New York. He is Professor Emeritus of Broadcasting. He served as president of the Lubec Historical Society from 2001 to 2012. The History Press published his book *Remembering Lubec: Stories from the Easternmost Point* in 2009.

Visit us at
www.historypress.net

This title is also available as an e-book